Internet Teletraffic Modeling and Estimation

RIVER PUBLISHERS SERIES IN INFORMATION SCIENCE AND TECHNOLOGY

Consulting Series Editor

KWANG-CHENG CHEN
National Taiwan University
Taiwan

Information science and technology enables 21st century into an Internet and multimedia era. Multimedia means the theory and application of filtering, coding, estimating, analyzing, detecting and recognizing, synthesizing, classifying, recording, and reproducing signals by digital and/or analog devices or techniques, while the scope of "signal" includes audio, video, speech, image, musical, multimedia, data/content, geophysical, sonar/radar, bio/medical, sensation, etc. Networking suggests transportation of such multimedia contents among nodes in communication and/or computer networks, to facilitate the ultimate Internet. Theory, technologies, protocols and standards, applications/services, practice and implementation of wired/wireless networking are all within the scope of this series. We further extend the scope for 21st century life through the knowledge in robotics, machine learning, cognitive science, pattern recognition, quantum/biological/molecular computation and information processing, and applications to health and society advance.

- Communication/Computer Networking Technologies and Applications
- Queuing Theory, Optimization, Operation Research, Statistical Theory and Applications
- Multimedia/Speech/Video Processing, Theory and Applications of Signal Processing
- Computation and Information Processing, Machine Intelligence, Cognitive Science, and Decision

For a list of other books in this series, please visit www.riverpublishers.com

Internet Teletraffic Modeling and Estimation

Alexandre Barbosa de Lima
Escola Politécnica of the University of São Paulo
São Paulo, Brazil

and

José Roberto de Almeida Amazonas
Escola Politécnica of the University of São Paulo
São Paulo, Brazil

River Publishers
Aalborg

ISBN 978-87-92982-10-0 (hardback)

Published, sold and distributed by:
River Publishers
P.O. Box 1657
Algade 42
9000 Aalborg
Denmark

Tel.: +45369953197
www.riverpublishers.com

Cover design by Fernando Freitas, BigHouse Web Development Bureau, Porto Alegre, Brazil

Contents

List of Tables ix

List of Figures xi

Preface xvii

List of acronyms and symbols xix

1 Introduction 1
- 1.1 Objectives of telecommunications carriers . . . 1
- 1.2 Traffic characteristics . . . 2
- 1.3 Questions and contributions . . . 3
- 1.4 Time series basic concepts . . . 3
  - 1.4.1 Time series examples . . . 3
  - 1.4.2 Operators notation . . . 5
  - 1.4.3 Stochastic processes . . . 8
  - 1.4.4 Time series modeling . . . 12

2 The fractal nature of network traffic 25
- 2.1 Fractals and self-similarity examples . . . 25
  - 2.1.1 The Hurst exponent . . . 27
  - 2.1.2 Sample mean variance . . . 31
- 2.2 Long range dependence . . . 33
  - 2.2.1 Aggregate process . . . 35
- 2.3 Self-similarity . . . 36
  - 2.3.1 Exact second order self-similarity . . . 37
  - 2.3.2 Impulsiveness . . . 37
- 2.4 Final remarks: why is the data networks traffic fractal? . . . 44

3 Modeling of long-range dependent teletraffic 47
- 3.1 Classes of modeling . . . 47

3.1.1 Non-parametric modeling 48
3.2 Wavelet transform 50
3.2.1 Multiresolution analysis and the discrete wavelet transform 54
3.3 Model MWM 65
3.4 Parametric modeling 69
3.4.1 ARFIMA model 69
3.4.2 ARFIMA models prediction - optimum estimation . 72
3.4.3 Forms of prediction 74
3.4.4 Confidence interval 75
3.4.5 ARFIMA prediction 76
3.5 Long memory statistical tests 76
3.5.1 R/S statistics 76
3.5.2 GPH test 77
3.6 Some H and d estimation methods 78
3.6.1 R/S statistics 78
3.6.2 Variance plot 78
3.6.3 Periodogram method 79
3.6.4 Whittle's method 80
3.6.5 Haslett and Raftery's MV approximate estimator . . 80
3.6.6 Abry and Veitch's wavelet estimator 80
3.7 Bi-spectrum and linearity test 83
3.8 KPSS stationarity test 88

4 State-space modeling **91**
4.1 Introduction 91
4.2 TARFIMA model 92
4.2.1 Multistep prediction with the Kalman filter 94
4.2.2 The prediction power of the TARFIMA model . . . 95
4.3 Series exploratory analysis 97
4.3.1 ARFIMA(0; 0.4; 0) series 98
4.3.2 MWM series with $H = 0.9$ 102
4.3.3 Nile river series 107
4.4 Prediction empirical study with the TARFIMA model 116
4.4.1 ARFIMA$(0, d, 0)$ series 116
4.4.2 MWM series 119
4.4.3 Nile river series between years 1007 and 1206 121
4.4.4 Conclusions 124

5 Modeling of Internet traffic **127**
5.1 Introduction 127
5.2 Modeling of the UNC02 trace 127
5.2.1 Exploratory analysis 127
5.2.2 Long memory local analysis of the UNC02 trace . . 146
5.2.3 Empirical prediction with the TARFIMA model . . . 147

6 Conclusions **151**

Bibliography **153**

Index **161**

About the Authors **165**

List of Tables

3.1 Asymptotic values of the shape parameter p of the β multipliers as function of α (or H). 69

4.1 P_h^2 values for AR(1) and TARFIMA(L), $L = \{1, 10\}$ models. 97

4.2 P_h^2 for TARFIMA(L), $L = \{100, 150\}$ and ARFIMA($0, d, 0$) (column in which $L = \infty$) models. 97

4.3 d parameter estimates for the ARFIMA(0; 0.4; 0) realization. 98

4.4 ARFIMA(0; 0.4; 0) series: stationarity, long memory, normality and linearity tests, and number of unit roots (column $\hat{m}$). 101

4.5 H estimates for the MWM($H = 0.9$) realization. 105

4.6 MWM($H = 0.9$) series: statistical tests results and number of unit roots. 105

4.7 Nile river series d parameter estimates. 111

4.8 Nile river series: stationarity, long memory, normality and linearity tests, and number of unit roots (column $\hat{m}$). 111

4.9 ARFIMA(0; 0.4; 0) series: $PCEMSE_{x,y}(t = 3996, K)$ of the AR(15), TARFIMA(10), TARFIMA(50) and TARFIMA(100) estimated models (columns in which $L = 10, 50$, and 100, respectively) relative to the AR(2) model. The TARFIMA(L) models used $\hat{d} = 0.3684$. 119

4.10 MWM($H = 0.9$) series: $PCEMSE_{x,y}(t = 3996, K)$ of the estimated AR(28), TARFIMA(10), TARFIMA(50) and TARFIMA(100) models relative to the AR(2). The TARFIMA(L) models used $\hat{d} = 0.3572$. 122

4.11 Nile river series: $PCEMSE_{x,y}(t = 1106, K)$ of the estimated AR(8), TARFIMA(10), TARFIMA(50) and TARFIMA(100) models relative to the AR(2). The TARFIMA(L) models used $\hat{d} = 0.4125$. 125

5.1 `UNC02bin1s` series: results of statistical tests (stationarity, long memory, normality and linearity) and quantification of the number of unit roots (column $\hat{m}$). 135
5.2 d parameter estimates for the `UNC02bin1s` series. 137
5.3 `UNC02bin1s` series ($\hat{d} = 0.3717$): $PCEMSE_{x,y}(t = 2700, K)$ of the estimated AR(18), TARFIMA(10), TARFIMA(50) and TARFIMA(100) models relative to the AR(2) model. 150

List of Figures

1.1 Johnson & Johnson quarterly earnings per share. 4
1.2 Global warming. 5
1.3 Speech signal: aaa...hhh. 6
1.4 New York stock exchange prices. 7
1.5 Earthquake and explosion seismic signals. 8
1.6 Box-Jenkins' iterative cycle. 13

2.1 Two instances of the Mandelbrot set. 26
2.2 The Sweden coast seen at different resolutions. 28
2.3 Three instances of a cauliflower showing its fractal nature. . 29
2.4 Five initial iterations of the Cantor set. 29
2.5 Drexel University's local area network Ethernet traffic. . . . 30
2.6 (l) Ethernet real traffic; (c) Poisson simulated traffic; (r) Self-similar simulated traffic. 31
2.7 Multifractal traffic simulated by means of the MWM model. 32
2.8 A LRD series' ACF with $H = 0.9$ and $N = 4096$ samples. . 33
2.9 SDF for same power AR(4) and FD(0.4) models. 34
2.10 Pareto I probability density functions with $x_m = 1$. 39
2.11 Pareto I probability distribution functions with $x_m = 1$. . . . 40
2.12 Realization of $S_\alpha(\sigma, \eta, \mu) = S_{1.2}(1, 1, 0)$ (256 samples). . . . 41
2.13 Symmetric stable distributions: density function. 42
2.14 Symmetric stable distributions: probability distribution. . . . 43
2.15 Asymmetric stable distributions: density function. 44
2.16 Asymmetric stable distributions: probability distribution. . . 45

3.1 Realizations of FBM processes for several Hurst parameters values. 48
3.2 Four examples of wavelet functions. 51
3.3 Meyer's wavelet. 52
3.4 Gaussian wavelet (related to the first derivative of a Gaussian PDF). 52

3.5 "Mexican hat" wavelet (related to the second derivative of a Gaussian PDF). 53
3.6 The image on the bottom part of the figure is the CWT $W_\psi(s, \tau)$ of the signal on the top part. 55
3.7 Critical sampling of the time-scale plane by means of the CWT parameters ($s = 2^j$ e $\tau = 2^j k$) discretization. 56
3.8 An illustration of the 3-levels DWT of the discrete time signal $x(k) = \sin(3k) + \sin(0.3k) + \sin(0.03k)$. 59
3.9 SDF of the signal $x(k) = \sin(3k) + \sin(0.3k) + \sin(0.03k)$. 59
3.10 Synthesis of the signal $x(k) = \sin(3k) + \sin(0.3k) + \sin(0.03k)$ in terms of the sum $\mathcal{S}_3(t) + \mathcal{D}_1(t) + \mathcal{D}_2(t) + \mathcal{D}_3(t)$. 60
3.11 QMF filters frequency response (graphs on the upper part) *vs* *brickwall*-type filters frequency response. 63
3.12 Daubechies' wavelets with $N = 2, 3, 4$ *vanishing moments*, bottom row from left to right, respectively. The corresponding scale functions are in the upper part. 64
3.13 QMF analysis filter banks $G^*(f)$ (low-pass) and $H^*(f)$ (high-pass) with decimation (*downsampling*) by a factor of 2. 65
3.14 QMF reconstruction filter banks with interpolation (*upsampling*) by a factor of 2. Observe that are used dual low-pass and high-pass filters, $G(f)$ and $H(f)$. 66
3.15 Flow diagram that shows the initial projection of a signal $x(t)$ on V_0 followed by the decomposition in W_1, W_2 and V_2. . . . 66
3.16 Flow diagram that illustrates the approximate synthesis of $x(t)$ from W_1, W_2 and V_2. 67
3.17 Block diagram that shows that the DWT works as a sub-bands codification scheme. The spectrum $U_0(f)$ of the signal $u_0(n)$ is subdivided in three frequency bands (that cover two octaves): $0 \leq f < 1/8$, $1/8 \leq f < 1/4$ and $1/4 \leq f \leq 1/2$. 67
3.18 SDFs for equal power AR(4) and FD(0,4). 71
3.19 From top to bottom, the wavelet spectra of a WGN, of the AR(4) $\boldsymbol{x}_t = 2,7607\boldsymbol{x}_{t-1} - 3,8106\boldsymbol{x}_{t-2} + 2,6535\boldsymbol{x}_{t-3} - 0,9238\boldsymbol{x}_{t-4} + \boldsymbol{w}_t$ model and of the `BellcoreAug89` trace (*bin* of 10 miliseconds). 84
3.20 Smoothed periodogram by the WOSA method: AR(4) 85
3.21 Smoothed periodogram by the WOSA method: Bellcore trace 85

4.1 Simulation: ARFIMA(0; 0.4; 0). 99

4.2 ARFIMA(0; 0.4; 0) simulation: periodogram. 100
4.3 ARFIMA(0; 0.4; 0) simulation: Daniell's window smoothed periodogram. 100
4.4 ARFIMA(0; 0.4; 0) simulation: estimated periodogram by the WOSA method. 101
4.5 Wavelet spectrum of the ARFIMA(0; 0.4; 0) signal. 101
4.6 QQ-plot. 102
4.7 Theoretical SACF and ACF graphs of the estimated ARFIMA(0; 0.3684; 0) and AR(15) models. 103
4.8 Residuals' QQ-plot. 103
4.9 Residuals' SACF graph. 104
4.10 AR(15) model's poles and zeros diagram. 104
4.11 Realization: MWM($H = 0.9$). 105
4.12 MWM($H = 0.9$) simulation: periodogram. 106
4.13 MWM($H = 0.9$) simulation: Daniell's window smoothed periodogram. 107
4.14 MWM($H = 0.9$) simulation: estimated periodogram by the WOSA method. 107
4.15 MWM($H = 0.9$) simulation: QQ-plot. 108
4.16 MWM($H = 0.9$) simulation: Theoretical SACF and ACFs graphs of the estimated ARFIMA(0; 0.3572; 0) (red line) and AR(28) (green line) models. 108
4.17 Residuals' QQ-plot. 109
4.18 Residuals' SACF graph. 109
4.19 AR(28) model's poles and zeros diagram. 110
4.20 Nile river series between years 1007 and 1206. 110
4.21 Nile river series: periodogram. 112
4.22 Nile river series: Daniell's window smoothed periodogram. . 112
4.23 Nile river series: estimated periodogram by the WOSA method. 113
4.24 Nile river series: wavelet spectrum. 113
4.25 Nile river series: QQ-plot. 114
4.26 Nile river series: theoretical SACF and ACF graphs of the estimated ARFIMA(0; 0.4125; 0) (red line) and AR(8) (green line) models. 114
4.27 ARFIMA(0; 0.4125; 0) model's residuals QQ-plot. 115
4.28 ARFIMA(0; 0.4125; 0) model's residuals SACF. 115
4.29 AR(8) model's poles and zeros diagram. 116

4.30 ARFIMA(0; 0.4; 0) series: observations, h-steps predictions, and 95% confidence intervals for the TARFIMA(100) and AR(15) models predictions. 117

4.31 ARFIMA(0; 0.4; 0) series: observations, h-steps predictions, and 95% confidence intervals for the TARFIMA(100) and AR(2) models predictions. 117

4.32 ARFIMA(0; 0.4; 0) series: TARFIMA100-AR15 - difference between the absolute prediction errors of the AR(15) and TARFIMA(100) models (graph $(\Delta_h(1) - \Delta_h(2))$ *vs.* t . . . 118

4.33 ARFIMA(0; 0.4; 0) series: TARFIMA100-AR2 - difference between the absolute prediction errors of the AR(2) and TARFIMA(100) models (graph $(\Delta_h(1) - \Delta_h(2))$ *vs.* t . . . 118

4.34 MWM($H = 0.9$) series: h-steps predictions, and 95% confidence intervals for the TARFIMA(100) and AR(28) models predictions *vs.* t 120

4.35 MWM($H = 0.9$) series: h-steps predictions, and 95% confidence intervals for the TARFIMA(100) and AR(2) models predictions *vs.* t 120

4.36 MWM($H = 0.9$) series: difference between of prediction errors absolute values of the AR(28) and TARFIMA(100) models 121

4.37 MWM($H = 0.9$) series: difference between of prediction errors absolute values of the AR(2) and TARFIMA(100) models 122

4.38 Nile river series: Série do rio Nilo: h-steps predictions, and 95% confidence intervals for the TARFIMA(100) and AR(8) models predictions *vs.* t 123

4.39 Nile river series: h-steps predictions, and 95% confidence intervals for the TARFIMA(100) and AR(2) models predictions *vs.* t 123

4.40 Nile river series: difference between of prediction errors absolute values of the AR(8) and TARFIMA(100) models . . 124

4.41 Nile river series: difference between of prediction errors absolute values of the AR(2) and TARFIMA(100) models . . 124

5.1 Inbound traffic (packets per bin count series) on an access Gigabit Ethernet link of UNC on 13/04/2002, time 19:30-21:30, 1 millisecond scale. 129

5.2 Inbound traffic (packets per bin count series) on an access Gigabit Ethernet link of UNC on 13/04/2002, time 19:30-21:30, 1 second scale. 129
5.3 QQ-plot of `UNC02Pktsbin1ms`. 130
5.4 QQ-plot of `UNC02Pktsbin1s`. 131
5.5 Fitting of a stable distribution with parameters $\{\alpha = 1.92, \sigma = 495, \eta = 0.86, \mu = 2.15 \times 10^4\}$, to the `UNC02Pktsbin1s` series 132
5.6 Wavelet spectrum of the `UNC02` trace (in packets/*bin*) estimated by means of the `UNC02Pktsbin1ms` series. 133
5.7 Wavelet spectrum of the `UNC02` trace (in packets/*bin*) estimated by means of the `UNC02Pktsbin1s` series. 134
5.8 Inbound traffic (bytes per bin count series) on an access Gigabit Ethernet link of UNC on 13/04/2002, time 19:30-21:30, 1 millisecond scale. 134
5.9 Inbound traffic (bytes per bin count series) on an access Gigabit Ethernet link of UNC on 13/04/2002, time 19:30-21:30, 1 second scale. 135
5.10 Fitting of a stable PDF with parameters $\{\alpha = 1.79, \sigma = 4.09 \times 10^5, \eta = 0.99, \mu = 6.35 \times 10^6\}$, to the `UNC02bin1s` series 136
5.11 Wavelet spectrum of the `UNC02` trace (in bytes/*bin*) estimated by means of the `UNC02Bytesbin1ms` series. 136
5.12 Wavelet spectrum of the `UNC02` trace (in bytes/*bin*) estimated by means of the `UNC02Bytesbin1s` series. 137
5.13 `UNC02bin1s` series: Daniell's window smoothed periodogram. 138
5.14 `UNC02bin1s` series: estimated periodogram by the WOSA method. 138
5.15 Série `UNC02bin1s`: theoretical SACF and ACFs graphs of the estimated AR(18) and ARFIMA(0; 0.3717; 1) models. . 139
5.16 Residuals of the estimated ARFIMA(0; 0.3717; 1) model for the `UNC02bin1s` series: QQ-plot. 140
5.17 Residuals of the estimated ARFIMA(0; 0.3717; 1) model for the `UNC02bin1s` series: SACF. 140
5.18 Residuals of the estimated ARFIMA(0; 0.3717; 1) model for the `UNC02bin1s` series: periodogram. 141
5.19 Residuals of the fitted TARFIMA(100) model for the `UNC02bin1s` series: periodogram. 142

5.20 Residuals of the fitted TARFIMA(100) model for the UNC02bin1s series: SACF. 142
5.21 Residuals of AR(18) model: periodogram. 143
5.22 Residuals of AR(18) model: SACF. 143
5.23 Residuals of AR(2) model: periodogram. 144
5.24 Residuals of AR(2) model: SACF. 145
5.25 Poles and zeros diagram of the AR(2) model. 145
5.26 Local analysis of the Hurst parameter of the UNC02bin1s series: 256 points window. 146
5.27 Local analysis of the Hurst parameter of the UNC02bin1s series: 512 points window. 147
5.28 Prediction of the UNC02bin1s series with origin at $t = 2700$ ($\hat{d} = 0.3717$): h-step predictions and 95% confidence intervals for the TARFIMA(100) and AR(18) models' predictions. 148
5.29 Prediction of the UNC02bin1s series with origin at $t = 2700$ ($\hat{d} = 0.3717$): h-step predictions and 95% confidence intervals for the TARFIMA(100) and AR(2) models' predictions. 148
5.30 Prediction of the UNC02bin1s series with origin at $t = 2700$ ($\hat{d} = 0.3717$): difference between the absolute prediction errors of the AR(18) and TARFIMA(100) models. . . . 149
5.31 Prediction of the UNC02bin1s series with origin at $t = 2700$ ($\hat{d} = 0.3717$): difference between the absolute prediction errors of the AR(2) and TARFIMA(100) models. 149

Preface

This book compiles the class notes of the Internet Teletraffic modeling and Estimation graduate course that has been delivered internationally for the first time in 2008 at the Universitat Politècnica de Catalunya and that has been upgraded since then.

The original model presented in this book, which is called TARFIMA, was developed by Prof. Alexandre B. Lima during his PhD program. The results obtained are very promising and new developments are being carried on.

Any doubts or suggestions for improvement can be forwarded to the authors: ablima@lcs.poli.usp.br, jra@lcs.poli.usp.br.

List of acronyms and symbols

ACF	Autocorrelation Function
AR	Auto Regressive
ARFIMA	Auto Regressive Fractionally Integrated and Moving Average
ARMA	Auto Regressive and Moving Average
ATM	Asynchronous Transfer Mode
CAC	Call Admission Control
CWT	Continuous Wavelet Transform
DFBM	Discrete-time Fractional Brownian Motion
DWT	Discrete Wavelet Transform
FBM	Fractional Brownian Motion
FGN	Fractional Gaussian Noise
FIR	Finite Impulse Response
FT	Fourier Transform
ICWT	Inverse Continuous Wavelet Transform
IDWT	Inverse Discrete Wavelet Transform
IETF	Internet Engineering Task Force
LAN	Local Area Network(s)
LRD	Long Range Dependence
MA	Moving Average

MRA	MultiResolution Analysis
MPLS	Multiprotocol Label Switching
MWM	Multifractal Wavelet Model
PDT	Packet Delay Transfer
PDV	Packet Delay Variation
PLR	Packet Loss Rate
PSTN	Public Switched Telephone Networks
QMF	Quadrature Mirrored Filters
QoE	Quality of Experience
QoS	Quality of Service
THRU	Throughput
UPC	Usage Parameter Control
WAN	Wide Area Network(s)
WFT	Windowed Fourier Transform
WGN	White Gaussian Noise
WIN	White Independent Noise
WN	White Noise

1

Introduction

This chapter introduces the motivation and the importance of studying traffic in data networks in order to ensure end-to-end Quality of Service (QoS) in any multimedia connection. This chapter reviews the principal concepts on stochastic processes, introduces time series and associated notation, and the Auto Regressive (AR), Moving Average (MA) and Auto Regressive and Moving Average (ARMA) models.

1.1 Objectives of telecommunications carriers

Presently the goals of telecommunications carriers may be summarized as the integration of voice, data, video, and image over a single packet switched convergent network. This network is also known as integrated services, multi-service or next generation network and presents the following requirements:

- acceptable performance for each kind of service;
- reduction of operational costs;
- flexibility to support current and future services;
- dynamic bandwidth allocation;
- integrated transport of all kinds of information;
- efficient usage of network resources by means of statistical multiplexing.

The backbone of this convergent network consists of a single IP infrastructure supporting QoS, virtual private networks and IP protocols v4 and v6, where QoS is a function or Packet Delay Transfer (PDT), Packet Delay Variation (PDV), Throughput (THRU) and Packet Loss Rate (PLR), i.e., $QoS = f(PDT, PDV, THRU, PLR)$.

The Internet Engineering Task Force (IETF) has proposed several technologies and standards for the implementation of QoS in the Internet such as:

- Multiprotocol Label Switching (MPLS) [27, 82, 106];
- constraint-based routing [27];
- traffic engineering;
- Integrated Services (IntServ) [17];
- Differentiated Services (DiffServ) [5, 13, 86, 121];

The implementation of mechanisms to monitor and control the teletraffic is a must for an efficient operation of convergent networks. Without traffic control, the unbounded demand of shared resources (buffers, bandwidth and processors) may seriously degrade the network performance. Traffic control is necessary to protect the QoS perceived by the users, Quality of Experience (QoE), and to ensure the efficient usage of network resources. Traffic control functions can be summarized as [29]:

a) Call Admission Control (CAC);
b) Usage Parameter Control (UPC);
c) Priority control;
d) Congestion control.

Functions a), c) and d) require online traffic prediction.

1.2 Traffic characteristics

Measurements [69, 94] have shown that aggregate network traffic has statistical properties quite different from the existing traffic in the Public Switched Telephone Networks (PSTN). Network traffic can be viewed as a byte counting series, i. e., quantities of bytes are transmitted at uniformly separated transmission intervals (time "bins").

A. K. Erlang showed in 1909 [34] that the number of generated telephone calls in a certain time interval can be modeled by the Poisson process [48]. In stark contrast, network traffic traces (series) have fractal properties such as Long Range Dependence (LRD) or self-similarity, and impulsiveness (great variability), at several time scales, which are not captured by the Poisson process. Network traffic in some Local Area Network(s) (LAN) and Wide Area Network(s) (WAN) is extremely impulsive because it has a heavy tail distribution [8, 40, 61–63, 99].

It is a well known fact that buffer overflow probability under Markovian traffic is an exponential function of the buffer size [64] (assume a single server who works at a constant rate service). However, an increase in buffer size produces a significant decrease in PLR. On the other hand, network traffic's long memory and impulsiveness degrade network performance (PLR increases) because buffer overflow probability in systems under fractal traffic is a power function of the buffer size, implying a hyperbolic decay much slower than the Markovian case [35, 46, 69, 99, 126]. Self-similar traffic produces buffer overflow much more frequently than Markovian traffic and such phenomena were observed in the early 1990s in Asynchronous Transfer Mode (ATM) networks implemented with small buffers (10 to 100 cells) [125].

1.3 Questions and contributions

This book reports the results obtained by Alexandre Barbosa de Lima during his PhD. His thesis proposed to answer the following questions:

- How can we apply the theory of time series to real network traffic?
- How can we identify an adequate model in practice?
- How can we make the estimation of the model's parameters?
- How can we make prediction of signals that present LRD?
- Should the prediction be made using a linear estimation theory, what are the available techniques?
- Is teletraffic a linear or non-linear signal? Is it Gaussian or non-Gaussian?

The contributions of his thesis may be summarized as:

- Development of a new fractal teletraffic model in the state space.
- Proposal of a LRD teletraffic statistical analysis methodology.
- Demonstration that modeling of signals with long memory by means of high-order AR processes may be unfeasible in practice.

1.4 Time series basic concepts

1.4.1 Time series examples

The following figures present some examples of time series obtained from different domains. Roughly speaking, time series are set of numbers corresponding to the observation of a certain phenomenon. By nature, such numbers are random variables.

Figure 1.1 shows the Johnson & Johnson quarterly earnings per share. It is possible to realize a continuous trend of increase.

Figure 1.2 shows the evolution of the global warming along the years. In this case, it is observed an oscillation around an average value till the years 1960 and from then a trend to increase.

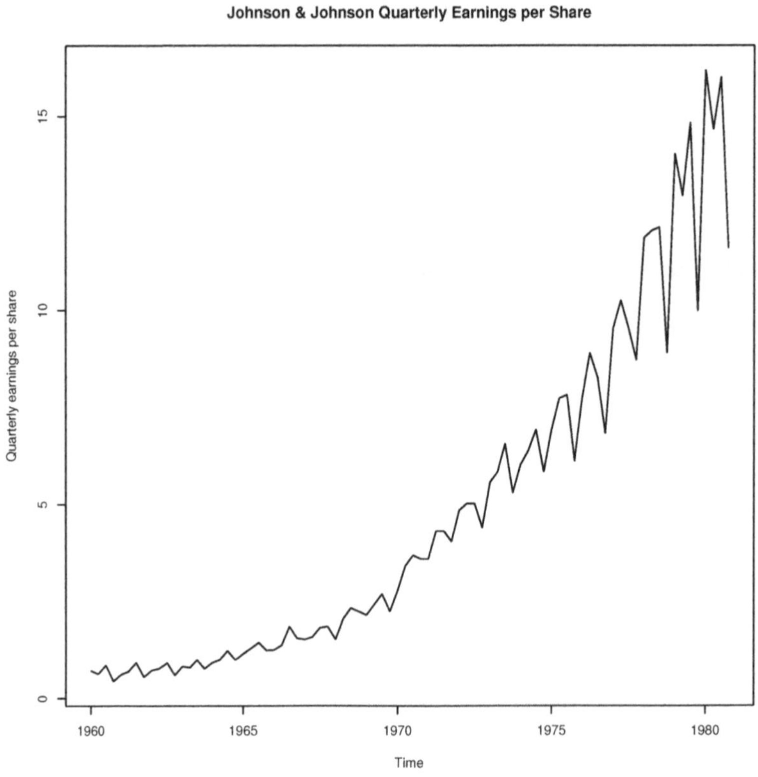

Figure 1.1: Johnson & Johnson quarterly earnings per share.

Figure 1.3 shows the speech signal corresponding to the vocalization of *aaa...hhh*. It is easy to observe a periodic behavior and a decrease of the power as the time increases.

Figure 1.4 shows the variation of the New York stock exchange prices. The distinctive feature is an abrupt variation around sample 800.

Figure 1.5 shows representative seismic signals of an earthquake and of an nuclear bomb explosion beneath the ground. Visually both signals are different, however what should be the difference in their mathematical structures that would allow to differentiate them by automatic means?

In all cases, the objective is to find models that allow to understand the physical phenomena and to estimate the future behavior. Accurate predictions or forecasts are power tools to help the decision make process as it may leverage the gains and minimize risks.

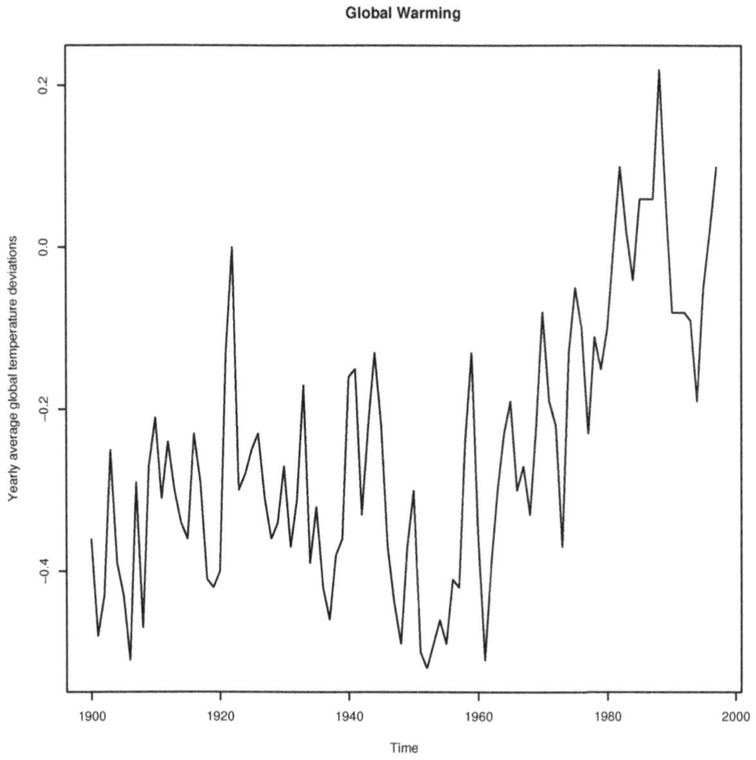

Figure 1.2: Global warming.

1.4.2 Operators notation

This section introduces the mathematical operators that are needed to represent and manipulate time series.

Consider a time series given by $\boldsymbol{x}_t$. Then:

(a) 1 sample delay operator, denoted by B:

$$B\boldsymbol{x}_t = \boldsymbol{x}_{t-1}. \tag{1.1}$$

(b) m samples delay operator, denoted by B^m, $m \in \mathbb{Z}$:

$$B^m\boldsymbol{x}_t = \boldsymbol{x}_{t-m}. \tag{1.2}$$

The impulse response h_t of the m samples delay system is [89]:

$$h_t = \delta_{t-m}, \tag{1.3}$$

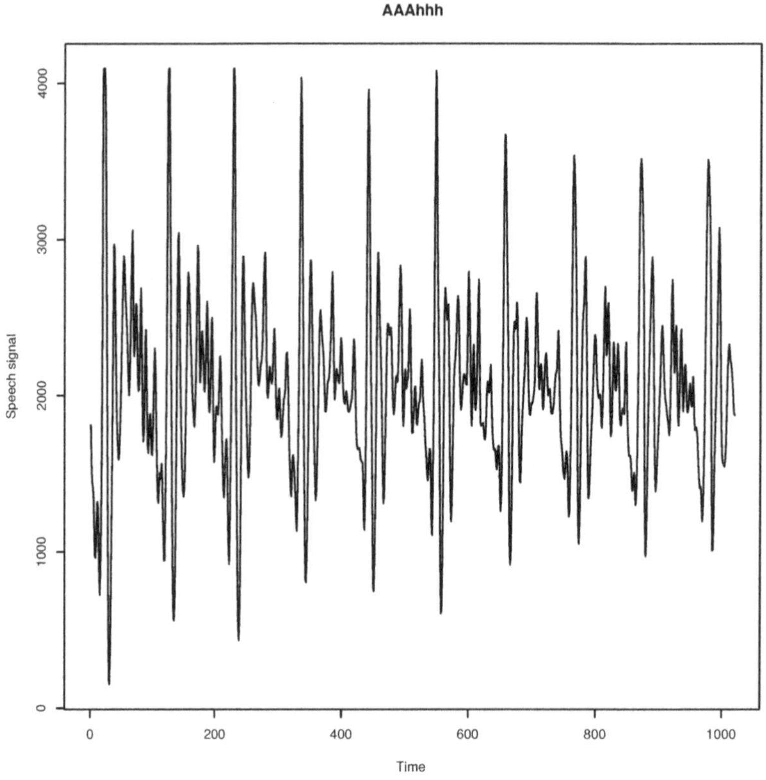

Figure 1.3: Speech signal: aaa...hhh.

and its transfer function, defined as the z transform($H(z) = \sum_{t=-\infty}^{\infty} h_t z^{-t}$) of the impulse response is:

$$H(z) = z^{-m}. \tag{1.4}$$

(c) difference or backshift operator

$$\Delta \boldsymbol{x}_t = \boldsymbol{x}_t - \boldsymbol{x}_{t-1} = (1 - B)\boldsymbol{x}_t, \tag{1.5}$$

So, it is possible to write, $B^j(\boldsymbol{x}_t) = \boldsymbol{x}_{t-j}$ and $\Delta^j(\boldsymbol{x}_t) = \Delta(\Delta^{j-1}(\boldsymbol{x}_t))$, $j \geq 1$, with $\Delta^0(\boldsymbol{x}_t) = \boldsymbol{x}_t$.
Accordingly, $\Delta^2 \boldsymbol{x}_t = \Delta(\Delta \boldsymbol{x}_t) = (1-B)(1-B)\boldsymbol{x}_t = (1-2B+B^2)\boldsymbol{x}_t = \boldsymbol{x}_t - 2\boldsymbol{x}_{t-1} + \boldsymbol{x}_{t-2}$.

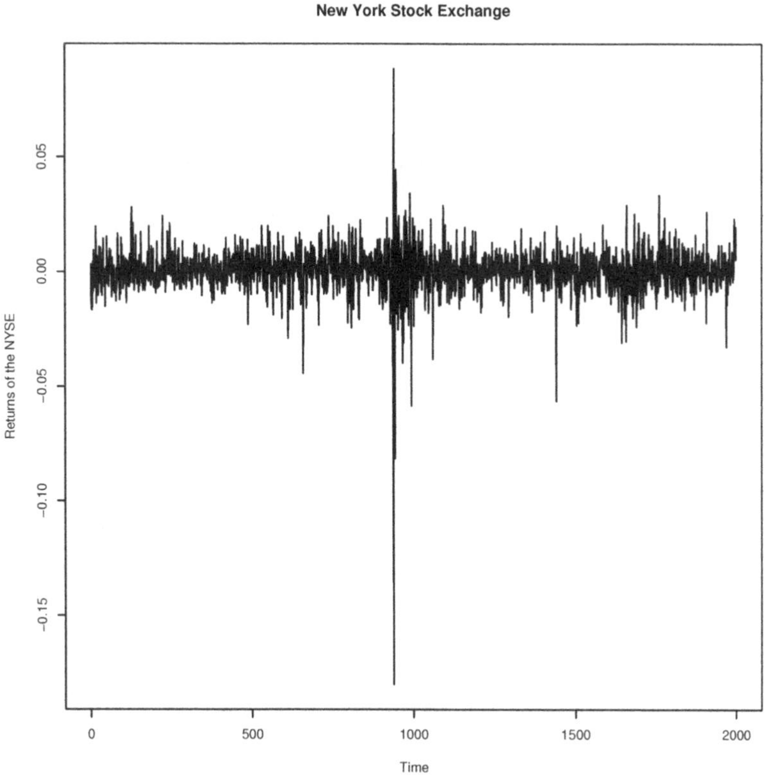

Figure 1.4: New York stock exchange prices.

The difference operator has impulse response:

$$h_t = \delta_t - \delta_{t-1}, \tag{1.6}$$

and transfer function

$$H(z) = (1 - z^{-1}). \tag{1.7}$$

(d) summation operator or integrator filter, denoted by S:

$$\begin{aligned} S\boldsymbol{x}_t &= \sum_{i=0}^{\infty} \boldsymbol{x}_{t-i} = \boldsymbol{x}_t + \boldsymbol{x}_{t-1} + \boldsymbol{x}_{t-2} + \ldots = \\ &= (1 + B + B^2 + \ldots)\boldsymbol{x}_t = (1 - B)^{-1}\boldsymbol{x}_t = \Delta^{-1}\boldsymbol{x}_t. \end{aligned} \tag{1.8}$$

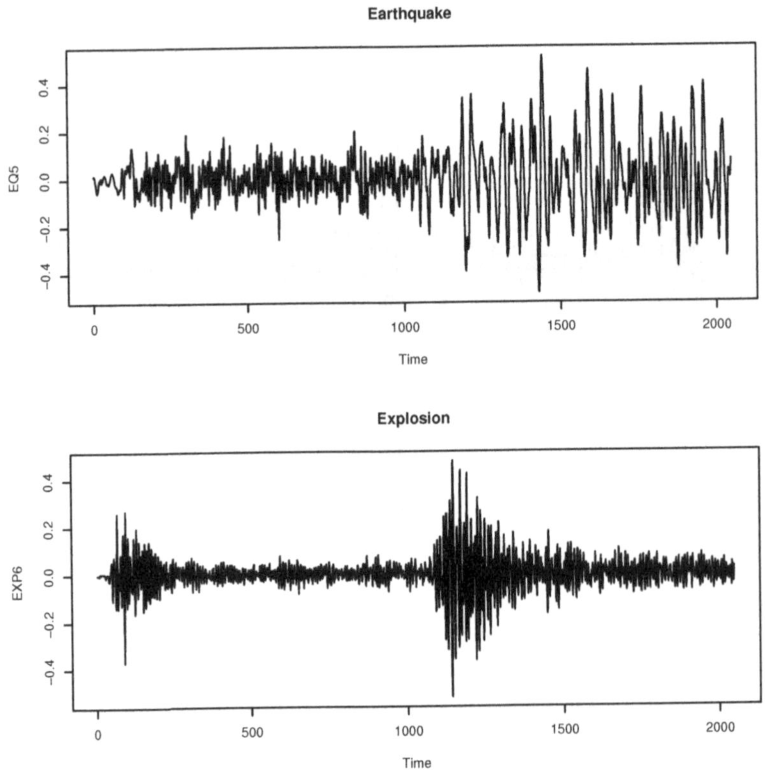

Figure 1.5: Earthquake and explosion seismic signals.

The integrator filter's transfer function corresponds to the inverse of the system's function defined by (1.7), i. e.,

$$H(z) = (1 - z^{-1})^{-1}. \tag{1.9}$$

1.4.3 Stochastic processes

This section introduces a very short review of the main concepts about stochastic processesthat are used in this book. The purpose is just to indicate what is important to be known without entering into the details. The interested reader should refer to the appropriate literature [15, 19, 20, 83, 96, 118, 128].

Definition 1.4.3.1 (Stochastic Process)**.** *Let T be an arbitrary set. A stochastic process is a family $\{\mathbf{x}_t, t \in T\}$, such that, for each $t \in T$, $\mathbf{x}_t$ is a random variable [85].* ■

When the set T is the set of integer numbers $\mathbb{Z}$, then $\{\boldsymbol{x}_t\}$ is a discrete time stochastic process (or random sequence); $\{\boldsymbol{x}_t\}$ is a continuous time stochastic process if T is taken as the set of real numbers $\mathbb{R}$.

The random variable $\boldsymbol{x}_t$ is, in fact, a function of two arguments $\boldsymbol{x}(t, \zeta)$, $t \in T$, $\zeta \in \Omega$, given that it is defined over the sample space Ω. For each $\zeta \in \Omega$ we have a realization, trajectory or time series x_t [83, p. 22]. The set of all realizations is called ensemble. Each trajectory is a function or a non-random sequence and for each fixed t, x_t is a number.

A process $\boldsymbol{x}_t$ is completely specified by its *finite-dimensional distributions* or n-order probability distribution functions, as:

$$F_x(x_1, x_2, \ldots, x_n; t_1, t_2, \ldots, t_n) = P\{\boldsymbol{x}(t_1) \leq x_1, \boldsymbol{x}(t_2) \leq x_2, \ldots, \boldsymbol{x}(t_n) \leq x_n\} \tag{1.10}$$

in which $t_1, t_2, \ldots, t_n$ are any elements of T and $n \geq 1$.

The first order probability distribution function is also known as Cumulative Distribution Function - CDF.

The *probability density function - PDF* is given by:

$$f_x(x_1, x_2, \ldots, x_n; t_1, t_2, \ldots, t_n) = \frac{\partial^n F_x(x_1, x_2, \ldots, x_n; t_1, t_2, \ldots, t_n)}{\partial x_1 \partial x_2 \ldots \partial x_n}. \tag{1.11}$$

Applying the conditional probability density formula,

$$f_x(x_k | x_{k-1}, \ldots, x_1) = \frac{f_x(x_1, \ldots, x_{k-1}, x_k)}{f_x(x_1, \ldots, x_{k-1})}, \tag{1.12}$$

in which $f_x(x_1, \ldots, x_{k-1}, x_k)$ denotes $f_x(x_1, \ldots, x_{k-1}, x_k; t_1, \ldots, t_{k-1}, t_k)$, repeatedly over $f_x(x_1, \ldots, x_{n-1}, x_n)$ we get the probability chain rule [114, p. 362]

$$f_x(x_1, x_2, \ldots, x_n) = f_x(x_1) f_x(x_2|x_1) f_x(x_3|x_2, x_1) \ldots f_x(x_n|x_{n-1}, \ldots, x_1). \tag{1.13}$$

When $\boldsymbol{x}_t$ is a sequence of *mutually independent* random variables, (1.13) can be rewritten as

$$f_x(x_1, x_2, \ldots, x_n) = f_x(x_1) f_x(x_2) \ldots f_x(x_n). \tag{1.14}$$

Definition 1.4.3.2 (Purely Stochastic Process). *A purely stochastic process $\{\mathbf{x}_t, t \in \mathbb{Z}\}$ is a sequence of mutually independent random variables.* ■

Definition 1.4.3.3 (IID Process). *An Independent and Identically Distributed (IID) process* $\{\mathbf{x}_t, t \in \mathbb{Z}\}$, *denoted by* $\mathbf{x}_t \sim IID$, *is a purely stochastic and identically distributed process.* ■

Definition 1.4.3.4 (Strict Sense Stationarity). *A random process* $\mathbf{x}_t$ *is stationary in the strict sense if [90, p. 297]*

$$F_{\mathbf{x}}(x_1, x_2, \ldots, x_n; t_1, t_2, \ldots, t_n) = F_{\mathbf{x}}(x_1, x_2, \ldots, x_n; t_1 + c, t_2 + c, \ldots, t_n + c), \text{for any } c. \quad ■ \tag{1.15}$$

The mean $\mu_x(t)$ of $\boldsymbol{x}_t$ is the expected value of the random variable $\boldsymbol{x}_t$:

$$\mu_x(t) = E\boldsymbol{x}_t = \int_{-\infty}^{\infty} x f_x(x; t) dx, \tag{1.16}$$

in which $f_x(x; t)$ is the first order probability density function of $\boldsymbol{x}_t$.

The autocorrelation $R_x(t_1, t_2)$ of $\boldsymbol{x}_t$ is the expected value of the product $\boldsymbol{x}_{t_1}\boldsymbol{x}_{t_2}$ [90, p. 288]:

$$R_x(t_1, t_2) = E\boldsymbol{x}_{t_1}\boldsymbol{x}_{t_2} = \int_{-\infty}^{\infty}\int_{-\infty}^{\infty} x_1 x_2 f_x(x_1, x_2; t_1, t_2) dx_1 dx_2, \tag{1.17}$$

in which $f_x(x_1, x_2; t_1, t_2)$ is the second order probability density function of $\boldsymbol{x}_t$.

The autocovariance $C_x(t_1, t_2)$ of $\boldsymbol{x}_t$ is the covariance of the random variables $\boldsymbol{x}_{t_1}$ and $\boldsymbol{x}_{t_2}$ [90, p. 289]:

$$C_x(t_1, t_2) = R_x(t_1, t_2) - \mu_{t_1}\mu_{t_2}. \tag{1.18}$$

The correlation coefficient $\rho_x(t_1, t_2)$ of $\boldsymbol{x}_t$ is the ratio:

$$\rho_x(t_1, t_2) = \frac{C_x(t_1, t_2)}{\sqrt{C_x(t_1, t_1) C_x(t_2, t_2)}}. \tag{1.19}$$

Many authors refer to $\rho_x(t_1, t_2)$ as Autocorrelation Function (ACF) [69, 83, 93, 94, 118, 128].

Definition 1.4.3.5 (Wide Sense Stationarity). *A random process* $\mathbf{x}_t$ *is wide sense stationary if its mean is constant [90, p. 298]*

$$E\mathbf{x}_t = \mu_{\mathbf{x}}, \tag{1.20}$$

and if its autocorrelation depends only on the lag $\tau = t_2 - t_1$*:*

$$R_{\mathbf{x}}(t_1, t_2) = R_{\mathbf{x}}(t_1, t_1 + \tau) = R_{\mathbf{x}}(\tau). \quad ■ \tag{1.21}$$

Observe that (1.21) implies an autocovariance that depends on the lag, i. e., $C_x(\tau)$. In this case, the stationary process' variance is constant and given by

$$\sigma_x^2 = C_x(0). \tag{1.22}$$

Real series generally present the following types of non-stationarity [83]:

(a) level-based non-stationarity: the series oscillate around an average level during a certain time and then jump to another level. The first difference makes these series stationary.
(b) slope-based non-stationarity: the series float around a straight line, with positive or negative slope. The second difference makes these series stationary.

Definition 1.4.3.6 (White Independent Noise). *A process $\mathbf{w}_t \sim IID$ is a White Independent Noise (WIN) when it has mean $\mu_{\mathbf{w}}$ and variance $\sigma^2_{\mathbf{w}}$, $\mathbf{w}_t \sim WIN(\mu_{\mathbf{w}}, \sigma^2_{\mathbf{w}})$.* ■

Definition 1.4.3.7 (White Noise). *A sequence $\{\mathbf{w}_t, t \in \mathbb{Z}\}$ of non-correlated with mean $\mu_{\mathbf{w}}$ and variance $\sigma^2_{\mathbf{w}}$ is called White Noise (WN), $\mathbf{w}_t \sim WN(\mu_{\mathbf{w}}, \sigma^2_{\mathbf{w}})$.* ■

Observation 1.4.3.1 (White Gaussian Noise). *A White Gaussian Noise (WGN) $\mathbf{w}_t$ is a WIN and denoted by $\mathbf{w}_t \sim \mathcal{N}(\mu_{\mathbf{w}}, \sigma^2_{\mathbf{w}})$, in which $\mathcal{N}$ stands for the probabilities of a normal (Gaussian) distribution.*

Consider a stationary process $\boldsymbol{x}_t$. The sample mean of a realization x_t with N points is given by

$$\bar{x} = \frac{1}{N}\sum_{t=1}^{N} x_t, \tag{1.23}$$

the sample autocovariance of lag τ by

$$\hat{C}_\tau = \frac{1}{N}\sum_{t=\tau+1}^{N} (x_t - \bar{x})(x_{t-\tau} - \bar{x}). \tag{1.24}$$

The sample autocorrelation (SACF) of lag τ by

$$\hat{\rho}_\tau = \frac{\hat{C}_\tau}{\hat{C}_0}, \tag{1.25}$$

in which $\hat{C}_0$ (also denoted as $s_{\mathbf{x}}^2$)

$$\hat{C}_0 = s_{\mathbf{x}}^2 = \frac{1}{N}\sum_{t=1}^{N}(x_t - \bar{x})^2 \tag{1.26}$$

is the sample variance of x_t.

The sample variance can also be defined as $s_{\mathbf{x}}^2 = \frac{1}{N-1}\sum_{t=1}^{N}(x_t - \bar{x})^2$.

Definition 1.4.3.8 (Ergodicity). *A stationary process $\mathbf{x}_t$ is called ergodic if its main moments converge in probability*[1] *to the population's moments, i. e., if $\bar{\mathbf{x}} \xrightarrow{P} \mu$, $\hat{C}_\tau \xrightarrow{P} C_\tau$ e $\hat{\rho}_\tau \xrightarrow{P} \rho_\tau$ [128, p. 58].* ■

For example, if $\boldsymbol{x}_t = A$ where A is a random variable, then $\boldsymbol{x}_t$ is not ergodic.

1.4.4 Time series modeling

A time series x_t modeling consists on estimating an invertible function $h(.)$, called *model* of $\boldsymbol{x}_t$, such that

$$\boldsymbol{x}_t = h(\ldots, \boldsymbol{w}_{t-2}, \boldsymbol{w}_{t-1}, \boldsymbol{w}_t, \boldsymbol{w}_{t+1}, \boldsymbol{w}_{t+2}, \ldots), \tag{1.27}$$

in which $\boldsymbol{w}_t \sim$ IID and

$$g(\ldots, \boldsymbol{x}_{t-2}, \boldsymbol{x}_{t-1}, \boldsymbol{x}_t, \boldsymbol{x}_{t+1}, \boldsymbol{x}_{t+2}, \ldots) = \boldsymbol{w}_t, \tag{1.28}$$

in which $g(.) = h^{-1}(.)$. The process $\boldsymbol{w}_t$ is the *innovation* at instant t and represents the new information about the series that is obtained at instant t.

In practice, the adjusted model is *causal*, i. e.,

$$\boldsymbol{x}_t = h(\boldsymbol{w}_t, \boldsymbol{w}_{t-1}, \boldsymbol{w}_{t-2}, \ldots). \tag{1.29}$$

The model construction methodology is based on the iterative cycle illustrated by the following steps [16, 83]:

(a) a model general class is considered for analysis (*specification*);

(b) there is the *identification* of a model, based on statistical criteria;

---

1 We say that the sequence $\{\boldsymbol{x}_1, \boldsymbol{x}_2, \ldots, \boldsymbol{x}_n, \ldots\}$ converges in probability to x if $\lim_{n\to\infty} P(|\boldsymbol{x}_n - x| \geq \epsilon) = 0$ for all $\epsilon > 0$.

(c) it follows the *estimation* phase, in which the model's parameters are obtained. In practice, it is important that the model is *parsimonious*[2] and

(d) at last, there is the *diagnostic* of the adjusted model by means of a statistical analysis of the series of residues w_t (is w_t compatible with a WN?)

Figure 1.6 illustrates the Box-Jenkins' iterative cycle used built an appropriate model for a time series.

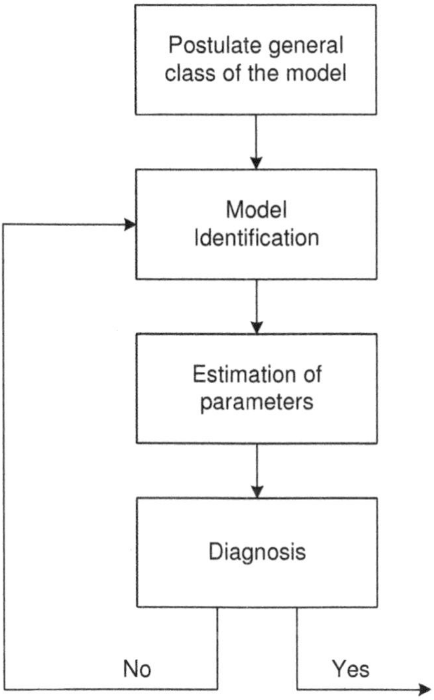

Figure 1.6: Box-Jenkins' iterative cycle.

[2] We say that a model is parsimonious when it uses few parameters. The use of an excessive number of parameters is undesirable because the uncertainty degree of the statistical inference procedure increases with the number of parameters.

The process $\boldsymbol{x}_t$ of (1.29) is *linear* when it corresponds to the convolution of a process $\boldsymbol{w}_t \sim \text{IID}$ and a deterministic sequence h_t [109, p. 377]

$$\begin{aligned}\boldsymbol{x}_t &= h_t \star \boldsymbol{w}_t = \sum_{k=0}^{\infty} h_k \boldsymbol{w}_{t-k} \\ &= \boldsymbol{w}_t + h_1 \boldsymbol{w}_{t-1} + h_2 \boldsymbol{w}_{t-2} + \ldots \\ &= (1 + h_1 B + h_2 B^2 + \ldots)\boldsymbol{w}_t \\ &= H(B)\boldsymbol{w}_t\end{aligned} \tag{1.30}$$

in which the symbol $\star$ denotes the convolution operation and $h_0 = 1$.

Eq. (1.30) is also known as the *infinite order moving average* (MA(∞)) representation [19].

The linear filter general format of (1.30) is

$$\boldsymbol{x}(t) = \sum_{k=1}^{p} \phi_k \boldsymbol{x}(t-k) + \boldsymbol{w}(t) - \sum_{k=1}^{q} \theta_k \boldsymbol{w}(t-k). \tag{1.31}$$

The sequence h_t is called impulse response of (1.31), also known as *ARMA model* of orders p and q (ARMA(p, q)).

In a more compact format, we have

$$\phi(B)\boldsymbol{x}_t = \theta(B)\boldsymbol{w}_t, \tag{1.32}$$

in which $\phi(B)$ is the order p auto-regressive operator

$$\phi(B) = 1 - \phi_1 B - \phi_2 B^2 - \ldots - \phi_p B^p \tag{1.33}$$

and $\theta(B)$ is the q order moving average operator

$$\theta(B) = 1 - \theta_1 B - \theta_2 B^2 - \ldots - \theta_q B^q. \tag{1.34}$$

If all poles and zeros of the transfer function

$$H(z) = \sum_{k=0}^{\infty} h_k z^{-k} = \frac{\theta(z)}{\phi(z)} \tag{1.35}$$

of the filter (1.32) are inside the unity radius circle

$$\phi(z) = 0, \quad |z| < 1 \tag{1.36}$$

$$\theta(z) = 0, \quad |z| < 1 \tag{1.37}$$

then the process $\boldsymbol{x}_t$ given by (1.30) is *stationary* (or non-explosive) and *invertible*, respectively [20, p. 537], [109, p. 377].

1.4.4.1 Stable distribution

A random variable X with α-stable distribution has a *heavy tail* and is defined by the following characteristic function [109]:

$$\Phi_X(w) = Ee^{jwX} = \int_{-\infty}^{\infty} f_X(x)e^{jwx}\,dx = \exp\{j\mu w - |\sigma w|^{\alpha} 1 - j\eta\,\mathrm{sign}(w)\varphi(w,\alpha)\}, \tag{1.38}$$

in which,

$$\varphi(w,\alpha) = \begin{cases} \tan(\alpha\pi/2) & \text{se } \alpha \neq 1 \\ -\frac{2}{\pi}\ln|w| & \text{se } \alpha = 1, \end{cases} \tag{1.39}$$

and sign(.) is the `sign` function, α $(0 < \alpha \leq 2)$ is the *characteristic exponent*, μ $(\mu \in \mathbb{R})$ is the *localization parameter*, η $(-1 \leq \eta \leq 1)$ is the *asymmetry parameter* and $\sigma \geq 0$ is the *dispersion parameter* or *scale*. The *variance* of X is *infinite* for $0 < \alpha < 2$.

When the innovations in (1.30) are random variables with α-stable distribution [37, 88, 109], (1.30) defines an *infinite variance non-Gaussian process* [20, p. 535], [109, p. 380].

This fact is justified by the *generalized central limit theorem* [88, 111], which states that, if the limit of an IID random variables sum converges, then this limit can only be a random variable with a stable distribution (the normal distribution is a particular case of the stable distribution[3]).

If the innovations in (1.30) have finite variance, i. e., $\boldsymbol{w}_t \sim \text{IWN}(\mu_w, \sigma_w^2)$, then $\boldsymbol{x}_t$ is Gaussian when h_t has infinite duration (central limit theorem) [114, p. 225]. Any non-linearity in the function $h(.)$ of (1.29) implies a *non-linear process* $\boldsymbol{x}_t$. In this case, $\boldsymbol{x}_t$ has statistics that are necessarily non-Gaussian. On the other hand, Gaussian processes are necessarily linear.

The rest of this Section assumes that the innovations in (1.30) are of the kind $\boldsymbol{w}_t \sim \text{WIN}(\mu_w, \sigma_w^2)$.

1.4.4.2 WIN's autocorrelation and SDF

The WIN's $\boldsymbol{w}_t$ autocorrelation in (1.30) is

$$R_{w(\tau)} = \sigma_w^2\delta_\tau + \mu_w^2, \tag{1.40}$$

in which δ_τ is the discrete time unit pulse[4].

[3] When $\alpha = 2$.

[4] $\delta_\tau = 1$ for $\tau = 0$, $\delta_\tau = 0$ para $\tau \neq 0$, $\tau \in \mathbb{Z}$.

Then, its spectral density function or SDF is

$$S_w(f) = \mu_w^2 \delta(f) + \sigma_w^2, \quad -1/2 \leq f \leq 1/2, \tag{1.41}$$

in which f is the normalized frequency and $\delta(f)$ is the Impulse generalized function (or Dirac's Delta).

The $\boldsymbol{w}_t$'s SDF is defined as the Discrete Time Fourier Transform (DFT) of its autocorrelation $R_{w(\tau)}$, i. e., $S_w(f) = \sum_{m=-\infty}^{\infty} R_{w(m)} e^{-j2\pi fm}$

1.4.4.3 The linear process $\boldsymbol{x}_t$'s SDF and auto-covariance

The linear process $\boldsymbol{x}_t$'s SDF is [89]

$$\begin{aligned} S_x(f) &= |H(f)|^2 S_w(f) \\ &= \mu_w^2 |H(0)|^2 \delta(f) + \sigma_w^2 |H(f)|^2, \quad -1/2 \leq f \leq 1/2, \end{aligned} \tag{1.42}$$

in which $H(f) = H(z)|_{z=e^{j2\pi f}}$ is the filter's frequency response.

The $\boldsymbol{x}_t$'s auto-covariance is given by [83, p. 112]

$$C_{x(\tau)} = \sigma_w^2 \sum_{t=0}^{\infty} h_t h_{t+\tau}, \tag{1.43}$$

and its variance by

$$\sigma_x^2 = \sigma_w^2 \sum_{t=0}^{\infty} h_t^2. \tag{1.44}$$

1.4.4.4 Auto-regressive models

As, in practice, the estimated models are invertible (i. e., (1.37) is valid), we can define the inverse operator $G(B) = H^{-1}(B)$ and rewrite (1.30) in the infinite order auto-regressive format (AR(∞))

$$\begin{aligned} \boldsymbol{x}_t &= g_1 \boldsymbol{x}_{t-1} + g_2 \boldsymbol{x}_{t-2} + \ldots + \boldsymbol{w}_t \\ &= \sum_{k=1}^{\infty} g_k \boldsymbol{x}_{t-k} + \boldsymbol{w}_t. \end{aligned} \tag{1.45}$$

So, $\boldsymbol{x}_t$ may be interpreted as a weighted sum of its past values $x_{t-1}, x_{t-2}, \ldots$ plus an innovation $\boldsymbol{w}_t$.

The equivalent model AR(∞) suggests that we can compute the probability of a future value $\boldsymbol{x}_{t+k}$ being between two specified bounds, i. e., (1.45) states that it is possible to make inferences or *predictions* of the series' future values.

An order p auto-regressive model satisfies the equation

$$\phi(B)\boldsymbol{x}_t = \boldsymbol{w}_t. \tag{1.46}$$

in which $\phi(B)$ is an order p polynomial.

Multiplying both sides of (1.46) by $\boldsymbol{x}_{t-k}$ and taking the expectation we get

$$E\boldsymbol{x}_t\boldsymbol{x}_{t-k} = \phi_1 E\boldsymbol{x}_{t-1}\boldsymbol{x}_{t-k} + \phi_2 E\boldsymbol{x}_{t-2}\boldsymbol{x}_{t-k} + \ldots + \phi_p E\boldsymbol{x}_{t-p}\boldsymbol{x}_{t-k} + E\boldsymbol{w}_t\boldsymbol{x}_{t-k},$$

as $\boldsymbol{x}_{t-k}$ does not depend on $\boldsymbol{w}_t$, but only on the noise up to instant $t-k$, that are not correlated with $\boldsymbol{w}_t$, then $E\boldsymbol{w}_t\boldsymbol{x}_{t-k} = 0$, $k > 0$, and

$$C_x(k) = \phi_1 C_x(k-1) + \phi_2 C_x(k-2) + \ldots + \phi_p C_x(k-p), \quad k > 0. \tag{1.47}$$

Dividing (1.47) by $C_x(0) = \sigma_x^2$, we get

$$\rho_x(k) = \phi_1 \rho_x(k-1) + \phi_2 \rho_x(k-2) + \ldots + \phi_p \rho_x(k-p), \quad k > 0, \tag{1.48}$$

or

$$\phi(B)\rho_x(k) = 0. \tag{1.49}$$

Let G_i^{-1}, $i = 1, \ldots, p$, be the roots of $\phi(B) = 0$ (model's *characteristic equation*) [118, p. 40]. Then, we can write

$$\phi(B) = \prod_{i=1}^{p}(1 - G_i B),$$

and it can be shown that the general solution of (1.49) is [83, p. 116]

$$\rho_x(k) = A_1 G_1^k + A_2 G_2^k + \cdots + A_p G_p^k, \tag{1.50}$$

in which the constants A_i, $i = 1, 2, \ldots, p$, are determined by initial conditions over $\rho_x(0), \rho_x(1), \ldots, \rho_x(p-1)$.

As the roots of $\phi(B) = 0$ must be out of unit radius circle, we must have $|G_k| < 1$, $k = 1, \ldots, p$. So, the ACF plot of an AR(p) process will normally show a mixture of damping sine and cosine patterns and exponential decays.

As $B = z^{-1}$, then Eqs. 1.36 and 1.37 may be written as $\phi(B) = 0$ for $|B| > 1$ and $\theta(B) = 0$ for $|B| > 1$.

For example, consider an AR(2) model

$$\boldsymbol{x}_t = \phi_1 \boldsymbol{x}_{t-1} + \phi_2 \boldsymbol{x}_{t-2} + \boldsymbol{w}_t$$

and its ACF, which satisfies the second order difference equation

$$\rho_k = \phi_1 \rho_{k-1} + \phi_2 \rho_{k-2}, \quad k > 0$$

with initial values $\rho_0 = 1$ e $\rho_1 = \frac{\phi_1}{1-\phi_2}$. From (1.50), the general solution of this equation is [16, pág.59]

$$\rho_k = \frac{G_1(1-G_2^2)G_1^k - G_2(1-G_1^2)G_2^k}{(G_1 - G_2)(1 + G_1 G_2)}.$$

1.4.4.5 AR models identification

In practice, the order p of an AR series is unknown and must be empirically specified. There are two approaches [118]:

i) use of the *Partial Auto-correlation Function (PACF)*;
ii) use of some *model's selection (identification) criterium.*

Let ϕ_{mi} be the i-th coefficient of an AR(m) process, so the last coefficient is ϕ_{mm}. Making $k = 1, \ldots, m$ in (1.48) (in the following, we adopt the simplified notation $\rho_x(k) = \rho_k$) and considering that $\rho_k = \rho_{-k}$ (ACF's even symmetry), we get the *Yule-Walker equations* [16, p.64], [83, p.134], [128, p.69]

$$\begin{aligned}
\rho_1 &= \phi_{m1} + \phi_{m2}\rho_1 + \ldots + \phi_{mm}\rho_{m-1}, \\
\rho_2 &= \phi_{m1}\rho_1 + \phi_{m2} + \ldots + \phi_{mm}\rho_{m-2}, \\
&\vdots \\
\rho_m &= \phi_{m1}\rho_{m-1} + \phi_{m2}\rho_{m-2} + \ldots + \phi_{mm},
\end{aligned} \tag{1.51}$$

that may be written in matrix format

$$\begin{bmatrix} 1 & \rho_1 & \cdots & \rho_{m-1} \\ \rho_1 & 1 & \cdots & \rho_{m-2} \\ \vdots & \vdots & \cdots & \vdots \\ \rho_{m-1} & \rho_{m-2} & \cdots & 1 \end{bmatrix} \begin{bmatrix} \phi_{m1} \\ \phi_{m2} \\ \vdots \\ \phi_{mm} \end{bmatrix} = \begin{bmatrix} \rho_1 \\ \rho_2 \\ \vdots \\ \rho_m \end{bmatrix} \tag{1.52}$$

or in its compact format

$$R_m \phi_m = \tilde{\rho}_m, \tag{1.53}$$

in which R_m is the order m autocorrelations matrix, ϕ_m the model's parameters vector and $\tilde{\rho}_m$ is the autocorrelations vector.

Solving (1.53) for $m = 1, 2, \ldots$, we get

$$\begin{aligned}
\phi_{11} &= \rho_1 \\
\phi_{22} &= \frac{\begin{vmatrix} 1 & \rho_1 \\ \rho_1 & \rho_2 \end{vmatrix}}{\begin{vmatrix} 1 & \rho_1 \\ \rho_1 & 1 \end{vmatrix}} \\
\phi_{33} &= \frac{\begin{vmatrix} 1 & \rho_1 & \rho_1 \\ \rho_1 & 1 & \rho_2 \\ \rho_2 & \rho_1 & \rho_3 \end{vmatrix}}{\begin{vmatrix} 1 & \rho_1 & \rho_2 \\ \rho_1 & 1 & \rho_1 \\ \rho_2 & \rho_1 & 1 \end{vmatrix}}
\end{aligned} \tag{1.54}$$

And, in general,

$$\phi_{mm} = \frac{|R_m^*|}{|R_m|}, \tag{1.55}$$

in which $|R_m|$ is the m order autocorrelations matrix' determinant and R_m^* is matrix R_m with the last column replaced by the autocorrelations vector.

The sequence $\{\phi_{mm}, m = 1, 2, \ldots\}$ is the PACF.

It can be shown that an AR(p) model has $\phi_{mm} \neq 0$ for $m \leq p$ and $\phi_{mm} = 0$ for $m > p$ [16].

The PACF may be estimated by adjusting the AR(m), $m = 1, 2, \ldots$ models sequence

$$\begin{aligned}
\boldsymbol{x}_t &= \phi_{11}\boldsymbol{x}_{t-1} + \boldsymbol{w}_{1t} \\
\boldsymbol{x}_t &= \phi_{21}\boldsymbol{x}_{t-1} + \phi_{22}\boldsymbol{x}_{t-2} + \boldsymbol{w}_{2t} \\
&\vdots \\
\boldsymbol{x}_t &= \phi_{m1}\boldsymbol{x}_{t-1} + \phi_{m2}\boldsymbol{x}_{t-2} + \ldots + \phi_{mm}\boldsymbol{x}_{t-m} + \boldsymbol{w}_{mt}, \\
&\vdots
\end{aligned} \tag{1.56}$$

by the least squares method [118, p. 40], [128, p. 70].

The sequence $\{\hat{\phi}_{mm}, m = 1, 2, \ldots\}$ is the Sample Partial Autocorrelation Function (SPACF).

The basic idea of an ARMA model selection criterium (or information criterion) is to choose the orders k and l that minimize the quantity [85]

$$P(k,l) = \ln \hat{\sigma}_{k,l}^2 + (k+l)\frac{C(N)}{N}, \tag{1.57}$$

in which $\hat{\sigma}_{k,l}^2$ is a residual variance estimate obtained by adjusting an ARMA(k,l) model to the N series observations.

$C(N)$ is a function of the series size.

The quantity $(k+l)\frac{C(N)}{N}$ is called penalty term and it increases when the number of parameters increases, while $\hat{\sigma}_{k,l}^2$ decreases.

Akaike proposed the information criterium [2, 3]

$$AIC(k,l) = \ln \hat{\sigma}_{k,l}^2 + \frac{2(k+l)}{N}, \tag{1.58}$$

known as AIC, in which $\hat{\sigma}_{k,l}^2$ is the maximum likelihood estimator of σ_w^2 for an ARMA(k,l) model.

Upper bounds K and L for k and l must be specified. Eq. (1.58) has to be evaluated for all possible (k,l) combinations with $0 \le k \le K$ $0 \le l \le L$. In general, K and L are functions of N, for example, $K = L = \ln N$ [85, p. 85].

For the case of AR(p) models, (1.58) reduces to

$$AIC(k) = \ln \hat{\sigma}_k^2 + \frac{2k}{N}, \quad k \le K. \tag{1.59}$$

Another criterium that is much used is the (Schwarz) *Bayesian Information Criteria* (BIC) [128, p. 77]

$$BIC(k,l) = \ln \hat{\sigma}_{k,l}^2 + \frac{\ln N}{N}(k+l). \tag{1.60}$$

For the case of AR(p) models, (1.60) reduces to

$$BIC(k) = \ln \hat{\sigma}_k^2 + \frac{k \ln N}{N}. \tag{1.61}$$

1.4.4.6 AR models estimation

Having identified the AR model's order p, we can go to the parameters estimation phase. The methods of moments, Least Squares and Maximum Likelihood may be used [83, 85, 96]. As, in general, the moments estimators are not good [83], statistical packages as `S-PLUS`, `E-VIEWS`, etc., use some Least Squares or Maximum Likelihood estimator.

1.4.4.7 Non-stationary series modeling

A non-stationary process has time dependent moments. Some common types of non-stationarity:

- time dependent mean
- time dependent variance

A process $\boldsymbol{y}_t$ is called stationary with respect to trends if it is of the type

$$\boldsymbol{y}_t = TD_t + \boldsymbol{x}_t, \tag{1.62}$$

in which TD_t denotes the term of deterministic trends (constant, trend, season) that depends on t and $\boldsymbol{x}_t$ is a stationary process.

For example, the process $\boldsymbol{y}_t$ given by

$$\begin{aligned} \boldsymbol{y}_t &= \mu + \delta t + \boldsymbol{x}_t, \quad \boldsymbol{x}_t = \phi \boldsymbol{x}_{t-1} + \boldsymbol{w}_t \\ \boldsymbol{y}_t - \mu - \delta t &= \phi(\boldsymbol{y}_{t-1} - \mu - \delta(t-1)) + \boldsymbol{w}_t \\ \boldsymbol{y}_t &= c + \beta t + \phi \boldsymbol{y}_{t-1} + \boldsymbol{w}_t \end{aligned} \tag{1.63}$$

in which $|\phi| < 1$, $c = \mu(1-\phi) + \delta$, $\beta = \delta(1-\phi)t$ and $\boldsymbol{w}_t$ is a WGN with null mean and power σ^2, is a stationary AR(1) process with respect to trends [128].

1.4.4.8 ARIMA model

If a process corresponds to the difference of order $d = 1, 2, \ldots$ of $\boldsymbol{x}_t$

$$\boldsymbol{y}_t = (1-B)^d \boldsymbol{x}_t = \Delta^d \boldsymbol{x}_t \tag{1.64}$$

is stationary, then $\boldsymbol{y}_t$ can be represented by an ARMA(p, q) model

$$\phi(B)\boldsymbol{y}_t = \theta(B)\boldsymbol{w}_t. \tag{1.65}$$

In this case,

$$\phi(B)\Delta^d \boldsymbol{x}_t = \theta(B)\boldsymbol{w}_t \tag{1.66}$$

is an ARIMA(p, d, q) model and we say that $\boldsymbol{x}_t$ is an "integral" of $\boldsymbol{y}_t$ [83] because

$$\boldsymbol{x}_t = S^d \boldsymbol{y}_t. \tag{1.67}$$

As the ARIMA(p, d, q) model

$$H(z) = \frac{\theta(z)}{\phi(z)(1-z^{-1})^d} \tag{1.68}$$

is marginally stable [102], as it has d roots on the unit circle, $\boldsymbol{x}_t$ of (1.66) is a *homogeneous non-stationary* process (meaning *non-explosive*) or having *unit roots* [83, 118, 128].

Observe that [83, p.139]:

(a) $d = 1$ corresponds to homogeneous non-stationary series with respect to the level (they oscillate around a mean level during a certain time and then jump to another temporary level);

(b) $d = 2$ corresponds to homogeneous non-stationary series with respect to the trend (they oscillate along a direction for a certain time and then change to another temporary direction).

The ARIMA model (1.66) may be represented in three ways:

(a) ARMA$(p+d, q)$ (similar to Eq. (1.31))

$$\boldsymbol{x}(t) = \sum_{k=1}^{p+d} \varphi_k \boldsymbol{x}(t-k) + \boldsymbol{w}(t) - \sum_{k=1}^{q} \theta_k \boldsymbol{w}(t-k); \tag{1.69}$$

(b) AR(∞) (inverted format), given by (1.45) or

(c) MA(∞), according to (1.30).

1.4.4.9 Random walk

Consider the model $\boldsymbol{y}_t \sim I(1)$

$$\boldsymbol{y}_t = \boldsymbol{y}_{t-1} + \boldsymbol{x}_t, \tag{1.70}$$

in which $\boldsymbol{x}_t$ is a stationary process. If we assume the initial condition y_0, (1.70) can be rewritten as an integrated sum

$$\boldsymbol{y}_t = \boldsymbol{y}_0 + \sum_{j=1}^{t} \boldsymbol{x}_j. \tag{1.71}$$

The integrated sum $\sum_{j=1}^{t} \boldsymbol{x}_j$ is called stochastic trend and it is denoted by TS_t. Observe that

$$TS_t = TS_{t-1} + \boldsymbol{x}_t, \tag{1.72}$$

in which $TS_0 = 0$.

If $\boldsymbol{x}_t \sim \mathcal{N}(0, \sigma_x^2)$ in (1.70), then $\boldsymbol{y}_t$ is known as *random walk*.

Including a constant in the right side of (1.70), we have a random walk with *drift*,

$$\boldsymbol{y}_t = \theta_0 + \boldsymbol{y}_{t-1} + \boldsymbol{x}_t. \tag{1.73}$$

Given the initial condition y_0, we can write

$$\begin{aligned} \boldsymbol{y}_t &= y_0 + \theta_0 t + \sum_{j=1}^{t} \boldsymbol{x}_j \\ &= TD_t + TS_t \end{aligned} \tag{1.74}$$

The mean, variance, autocovariance and ACF of $\boldsymbol{y}_t$ are given by [84]

$$\mu_t = y_0 + t\theta_0 \tag{1.75}$$

$$\sigma^2(t) = t\sigma_{\boldsymbol{x}}^2 \tag{1.76}$$

$$C_k(t) = (t-k)\sigma_{\boldsymbol{x}}^2 \tag{1.77}$$

$$\rho_k(t) = \frac{t-k}{t}. \tag{1.78}$$

Observe that $\rho_k(t) \approx 1$ when $t >> k$ and the literature states that the random walk has "*strong memory*" [118].

The random walk's SACF decays linearly for large *lags*.

2

The fractal nature of network traffic

This chapter introduces the concept of fractals, presents the Hurst exponent as the most important parameter to characterize self-similarity and LRD which are directly related to the traffic impulsiveness.

2.1 Fractals and self-similarity examples

The shapes of classical geometry - triangles, circles, spheres, etc. - loose their structures when magnified. For example, a person on the Earth's surface has the impression that it is flat. On the other hand, an astronaut in orbit sees a round Earth. Suppose that someone has not been informed of being on a point of a circle which has a radius of hundreds of kilometers. This observer realizes the circle as a straight line even though it is not true.

Benoit B. Mandelbrot proposed in 1975 the term *fractal* (from latin *fractus*, that means fractured, broken) to describe mathematical objects having a details rich structure along several observation scales [78]. The Mandelbrot set is a mathematical fractal with a detailed structure (i. e., highly irregular) along an infinite series of scales. Figure 2.1 shows two instances of the Mandelbrot set. It is easy to realize that despite of the different scales, both figures are essentially the same.

Broadly speaking, as far as *Self-similarity* is concerned one may say that a self-similar object contains smaller copies of itself in all scales. The Mandelbrot set is a deterministic fractal as it is exactly self-similar.

Physical and human sciences provide several examples of random fractals, in which self-similarity occurs in a statistical sense [9, 38, 97]:

- climatology and hydrology time series;
- functional magnetic resonance of the human brain;
- fluids turbulent movements;
- financial data series;

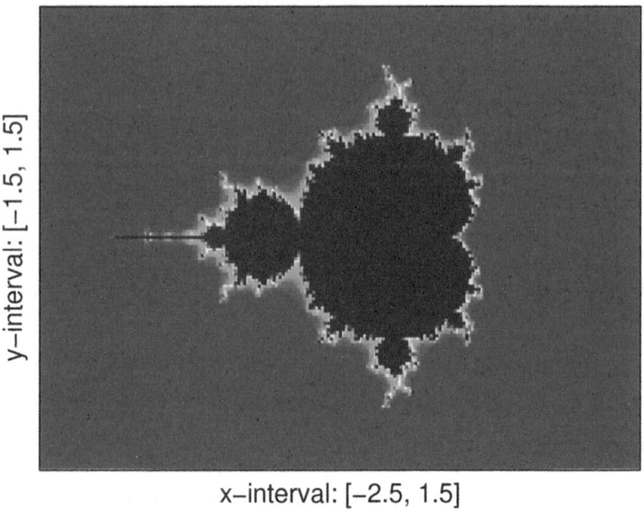

(a)

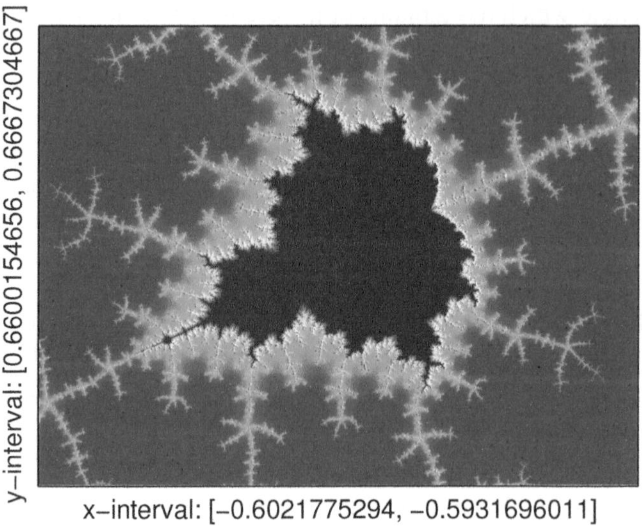

(b)

Figure 2.1: Fig. (a) is the Mandelbrot set. Fig (b) shows a Mandelbrot subset that is embedded in Fig. (a). The figures were obtained via `MATLAB` using a code developed by A. Klimke [65], from the University of Stuttgart.

- Sweden coast;
- cauliflower.

Figure 2.2 shows three pictures of the Sweden coast taken at different resolutions. From top to bottom the resolutions are 1.4 km, 500 m and 250 m. It is clearly seen that the same pattern repeats itself on all pictures.

Figure 2.3 presents three instances of a cauliflower showing its fractal nature.

Figure 2.4 shows the five initial iterations of the Cantor set, which is another example of a deterministic fractal [21] apud [95].

Figure 2.5 shows the Ethernet traffic collected at the Drexel University's local network. It is self-similar and highly impulsive at four aggregation time scales: (10 ms, 100 ms, 1 s and 10 s). The 100 ms scale series was obtained by aggregating the 10 ms scale series, i. e., a point at the 100 ms scale corresponds to the bytes addition in 10 consecutive bins of the 10 ms scale. It is surprising that successive aggregations do not smooth the traffic. Smoothing would happen if the traffic could be well modeled by the Poisson process [94].

Figure 2.6 shows on the left side an example of Ethernet real traffic [69]. It is possible to observe the self-similarity nature of the traffic, given that its impulsive nature at all scales. The central column shows a Poisson simulated traffic. In this case, aggregation smooths the traffic and does not correspond to what is observed in practice in a LAN. The right column shows self-similar simulated traffic that reproduces the behavior observed in practice.

2.1.1 The Hurst exponent

Due to historical reasons, the persistence degree (LRD) of a time series is characterized by means of the Hurst parameter $H, 0 < H < 1$ [54].

A time series is LRD (self-similar) when $1/2 < H < 1$ [9]. It is SRD (Short Range Dependence) when $0 < H \leq 1/2$. The closer is H to 1, the higher is the series persistence degree. We say the series is monofractal if H is time invariant. We say the series is multifractal, if H varies in time, both in a deterministic or random way.

It has been shown that the WAN traffic may be multifractal, with non-Gaussian marginal distribution, at refined time scales [36, 43, 103, 104]. On the other hand, a monofractal behavior has been observed for the LAN traffic [69]. Figure 2.7 shows a realization produced by the Riedi's MWM model [105].

Figure 2.8 illustrates the ACF of a long-memory process [28, 31]. The distinctive characteristic is its slow decay when compared to the autocorre-

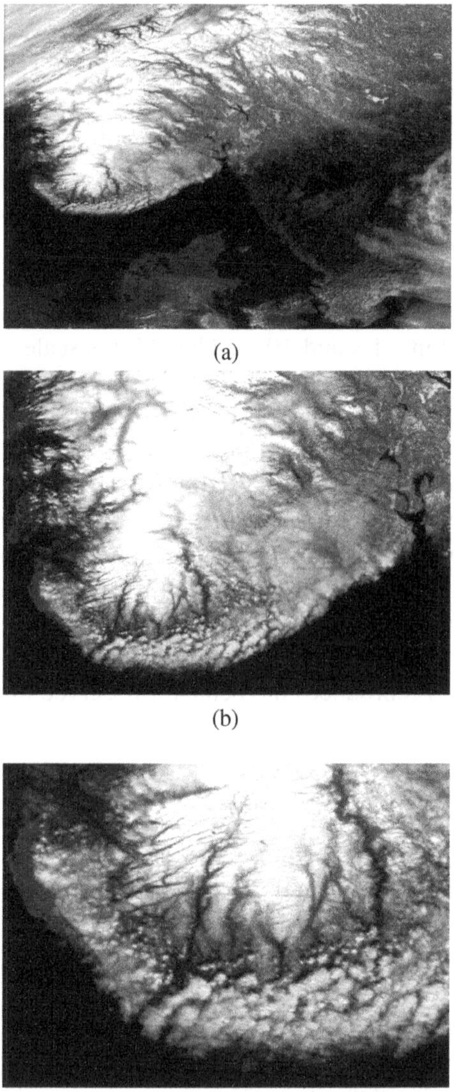

Figure 2.2: The Sweden coast seen at different resolutions. From top to bottom the resolutions are 1.4 km, 500 m and 250 m.

Figure 2.3: Three instances of a cauliflower showing its fractal nature.

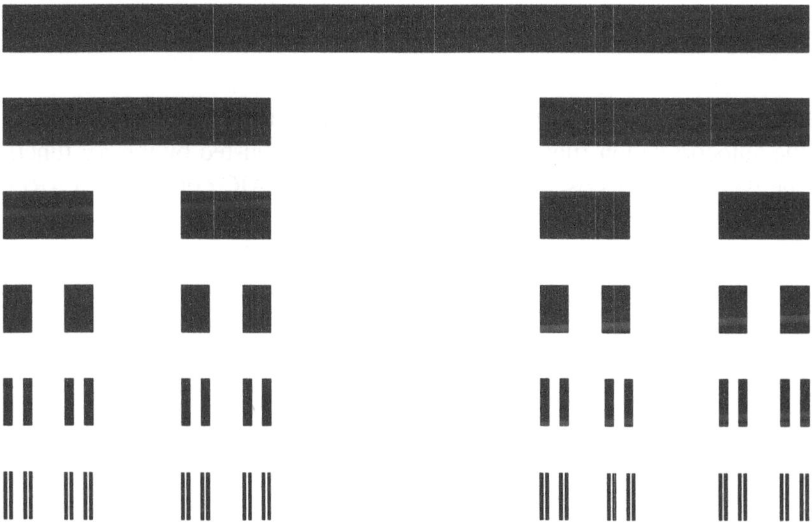

Figure 2.4: Five initial iterations of the Cantor set.

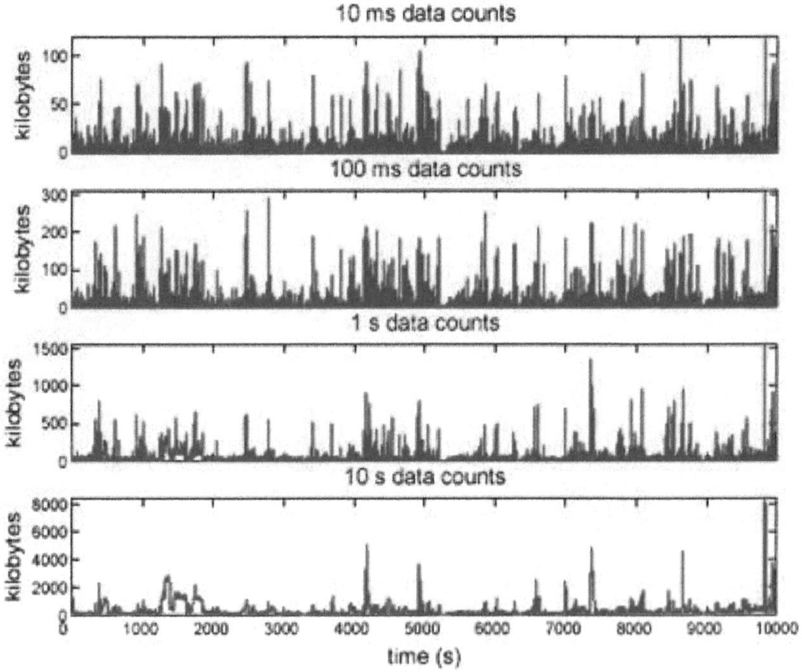

Figure 2.5: Drexel University's local area network Ethernet traffic.

lation function of the AR model that could be adjusted. The continuous line is the autocorrelation function of the AR model adjusted by the `ar` function of the `S+FinMetrics` software according to the AIC criterion. An AR(15) model was obtained.

Aggregate traffic can also show an ACF with long and short memory mixed characteristics [32, 73, 93]. This behavior is typical of the ARFIMA model class [45, 53]. Long memory is characterized in the frequency domain by an $1/f^{\alpha}singularity$, $0 < \alpha < 1$ ($\alpha = 2H - 1$), for $f \to 0$. Figure 2.9 shows that the SDF of a FD model class with $d = 0.4$ ($d = H - 1/2$) has an $1/f^{\alpha}$ behavior at the spectrum origin, while the AR(4) has not [45, 53]. FD(d) corresponds to an ARFIMA(p, d, q) with $p = q = 0$. The literature refers the LRD processes as $1/f$ noise.

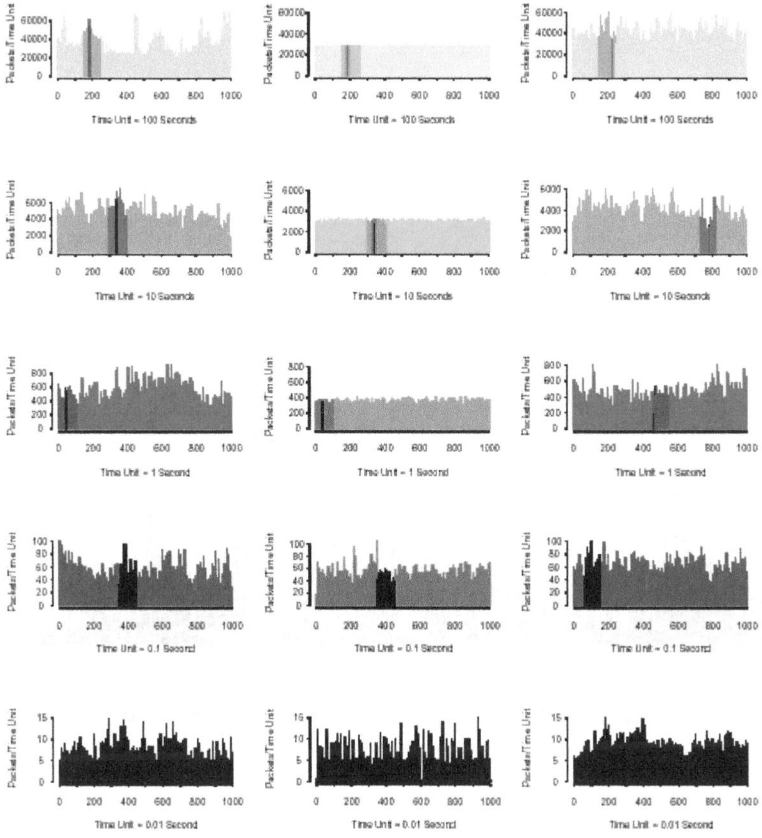

Figure 2.6: (l) Ethernet real traffic; (c) Poisson simulated traffic; (r) Self-similar simulated traffic.

2.1.2 Sample mean variance

Consider a stationary random process $\boldsymbol{x}_t$, $t \in \mathbb{Z}$, with mean $\mu_{\boldsymbol{x}}$ and variance $\sigma_{\boldsymbol{x}}^2$. Let $x_1, x_2, \ldots, x_N$ be the observations of a realization of x_t. If the random variables $\boldsymbol{x}_1, \boldsymbol{x}_2, \ldots, \boldsymbol{x}_N$ are independent or uncorrelated, then, the sample mean $\bar{x}$ variance is given by

$$\sigma_{\bar{x}}^2 = \frac{\sigma_{\boldsymbol{x}}^2}{N}. \tag{2.1}$$

If the sample is large enough, the $\bar{x}$ estimator's sample distribution is normal. The expression for the $\mu_{\boldsymbol{x}}$ confidence interval, at $(1 - \beta)$ confidence

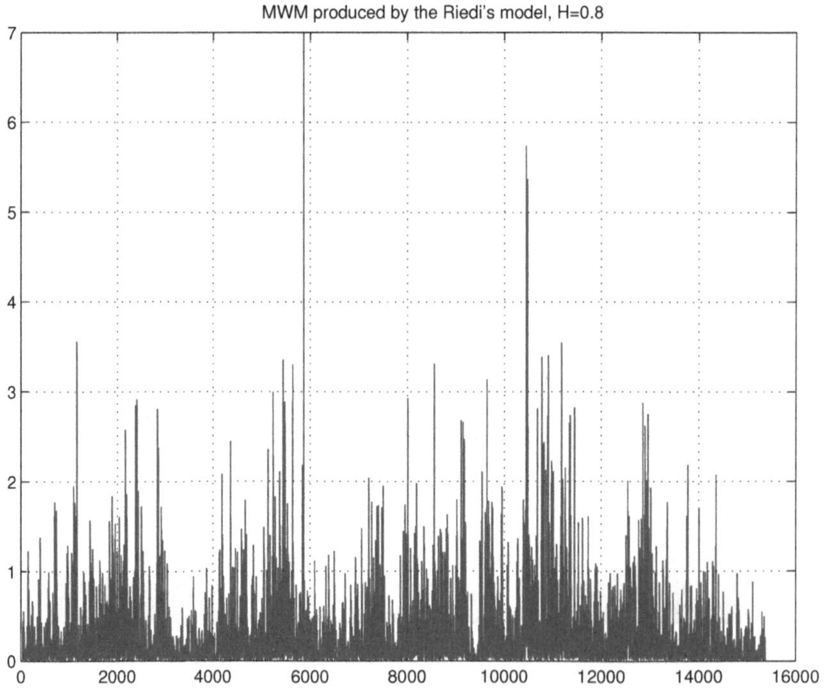

Figure 2.7: Multifractal traffic simulated by means of the MWM model. The time series impulsiveness varies in time: heteroscedasticity.

level is given by

$$\bar{x} - z_{\beta/2}\frac{\sigma_x}{\sqrt{N}} \leq \mu_x \leq \bar{x} + z_{\beta/2}\frac{\sigma_x}{\sqrt{N}}, \tag{2.2}$$

in which $z_{\beta/2}$ denotes the quantile $q_{(1-\beta/2)}$ of the standard normal distribution [9].

The quantile q_α of a distribution function F_x is the value for which $F_{q_\alpha} = \alpha$ [57, p.181]. The median, for example, corresponds to $q_{0,5}$. Given a probability $1 - \beta$, we find $z_{\beta/2}$ for which $P\{-z_{\beta/2} < Z < z_{\beta/2}\} = 1 - \beta$ ($z_{\beta/2} = 1,96$ para $1 - \beta = 95\%$).

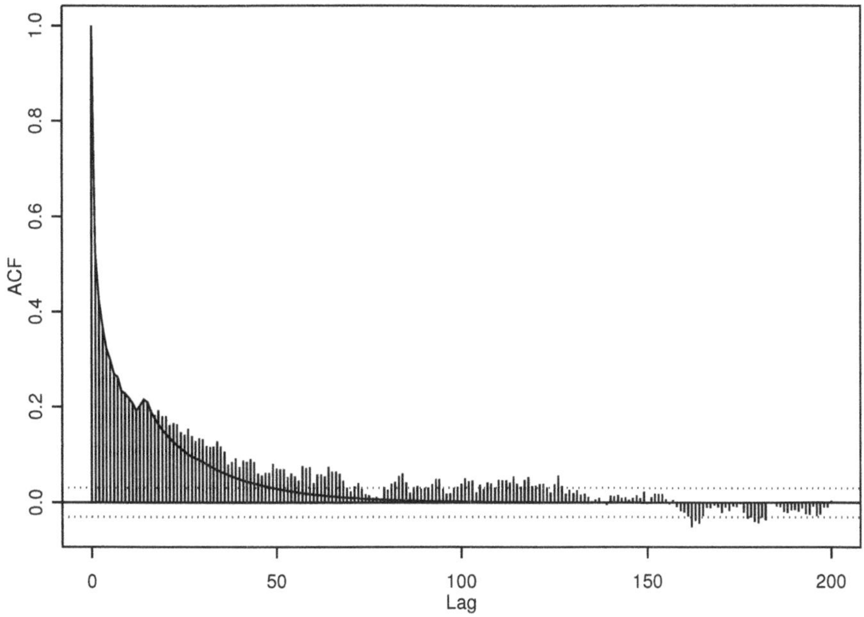

Figure 2.8: A LRD series' ACF with $H = 0.9$ and $N = 4096$ samples.

2.2 Long range dependence

Definition 2.2.0.1 (Long range dependence). $\mathbf{x}_t$ *is a long range dependence or long memory process if there are constants* α *and* C_P*, satisfying* $0 < \alpha < 1$ *and* $C_P > 0$*, for which [9, 97, 117]*

$$\lim_{f \to 0} \frac{P_{\mathbf{x}}(f)}{C_P |f|^{-\alpha}} = 1 , \tag{2.3}$$

where $P_{\mathbf{x}}(f)$ *denotes the SDF of* $\mathbf{x}_t$ *and* f *represents the normalized frequency* $(-1/2 \leq f \leq 1/2)$*, in cycles/sample.*

This is an asymptotic definition because the SDF is not specified for frequencies far from the origin.

The Hurst parameter and α are related by

$$H = \frac{\alpha + 1}{2}, \qquad 1/2 < H < 1 . \tag{2.4}$$

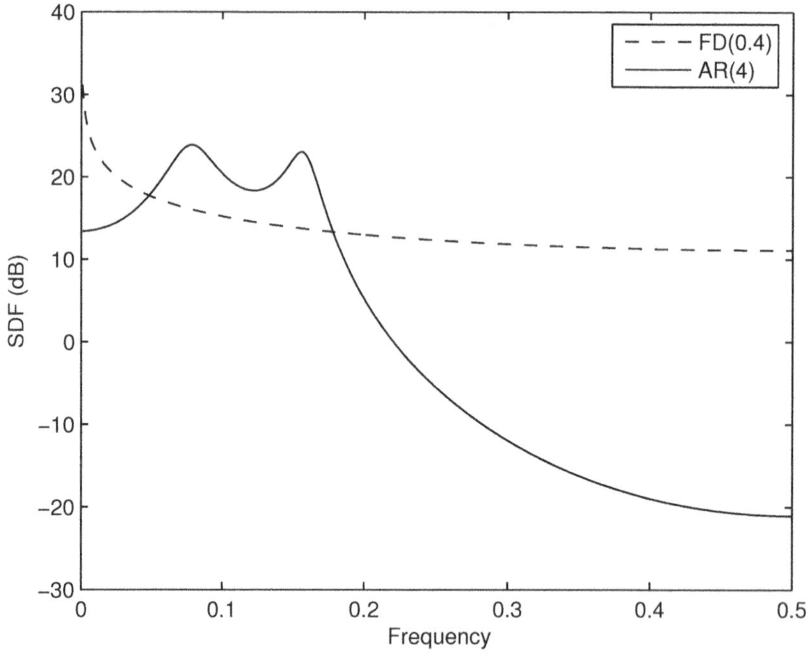

Figure 2.9: SDF for same power AR(4) and FD(0.4) models.

An alternate definition may be given in the time domain. $\boldsymbol{x}_t$ is a $1/f^{\alpha}$ process if its autocorrelation $R_x(\tau)$, for sufficiently large values of the *lag* τ, decreases according to a power function:

$$\lim_{\tau \to \infty} \frac{R_x(\tau)}{C_R \tau^{-(1-\alpha)}} = 1\,, \tag{2.5}$$

in which $C_R > 0$.

The singularity at the spectrum's origin implies

$$\sum_{\tau=-\infty}^{\infty} R_x(\tau) = \infty\,, \tag{2.6}$$

for $1/2 < H < 1$, i. e., the autocorrelations decay towards zero so slowly that they are not summable.

This behavior is drastically different from that presented by an ARMA process [15, 118] in which the autocorrelation's decay is fast, i. e., exponential:

$$|R_x(\tau)| \leq C r^{-\tau}, \quad \tau = 1, 2, \ldots\,, \tag{2.7}$$

in which $C > 0$ and $0 < r < 1$.

ARMA processes have short range dependence.

If $\boldsymbol{x}_t$ is LRD, the variance of $\bar{x}$ decreases with the sample size N more slowly than in the traditional case (independent or uncorrelated variables) [9, p.6]:

$$\sigma_{\bar{x}}^2 \approx \sigma_{\boldsymbol{x}}^2 c(\rho_x) N^{\alpha-1}, \tag{2.8}$$

in which $c(\rho_x)$ is defined by

$$\lim_{N\to\infty} N^{-(1+\alpha)} \sum_{i \neq j} \rho_x(i, j). \tag{2.9}$$

In this case, the $\bar{x}$ distribution is asymptotically Gaussian with $E\bar{x} = \mu_x$.

The LRD behavior of x_t makes parameters estimation, as $\bar{x}$, more difficult than that of uncorrelated observations. In this case, the confidence interval equation for μ_x (given by (2.2)) is no longer valid. In fact, for a given confidence level $(1 - \beta)$, the confidence interval must be "stretched" multiplying it by a factor F:

$$F = N^{\alpha/2}\sqrt{c(\rho_x)}. \tag{2.10}$$

This correction factor F increases with N, and in the limit goes to infinity.

2.2.1 Aggregate process

The M order aggregate process of $\boldsymbol{x}_t$, denoted by $X_t^{(M)}$, corresponds to a moving average of M size blocks without overlap of $\boldsymbol{x}_t$, i. e.,

$$X_i^{(M)} = \frac{1}{M} \sum_{t=M(i-1)+1}^{Mi} \boldsymbol{x}_t. \tag{2.11}$$

The following property is valid for a long memory process $\boldsymbol{x}_t$ [99]:

$$\lim_{M\to\infty} \frac{\operatorname{Var} X_t^{(M)}}{M^{(2H-2)}} = c, \tag{2.12}$$

in which c is a constant.

Aggregation is equivalent to a time scale change. It can be realized that the aggregate process is statistically similar to the original process, in the sense that a finite number of successive aggregations does not destroy the original process' impulsive character. Therefore, (2.12) suggests that long range dependence and self-similarity properties are closely related.

2.3 Self-similarity

Definition 2.3.0.1 (Processo H-ss). *A stochastic process* $\{\mathbf{y}_t\}_{t\in\mathbb{R}}$ *is self-similar with parameter* $0 < H < 1$*, i. e., it is* H-ss *if, for any* $a > 0$,

$$\{\mathbf{y}(t)\} \stackrel{d}{=} \{a^{-H}\mathbf{y}(at)\}, \tag{2.13}$$

in which $\stackrel{d}{=}$ *denotes equality between the finite-dimensions distributions [117].*

A H-ss process is LRD if $1/2 < H < 1$.

The Brownian movement (continuous time), also known as Wiener's process, is self-similar with $H = 1/2$ (but it is not LRD) [79].

If the process $\boldsymbol{x}_t = \Delta\boldsymbol{y}_t$, called process of $\boldsymbol{y}_t$ increments or first difference of $\boldsymbol{y}_t$, is stationary, then $\boldsymbol{y}_t$ is called H-sssi (H *self-similar with stationary increments*). In this case, the process H-sssi $\boldsymbol{y}_t$ is a first order integrated process, $\boldsymbol{y}_t \sim I(1)$.

If the moments of $\boldsymbol{y}(t)$ of order lower or equal to q exist, from (2.13) we conclude that [99]

$$E|\boldsymbol{y}_t|^q = E|\boldsymbol{y}_1|^q|t|^{qH}. \tag{2.14}$$

So, the process $\boldsymbol{y}_t \sim I(1)$ cannot be stationary.

Assuming $E\boldsymbol{y}_t = 0$ for the sake of simplifying notation[1], it can be shown that the $\boldsymbol{y}_t$ autocovariance is given by [9]

$$C_y(t,s) = E\boldsymbol{y}_t\boldsymbol{y}_s = \frac{\sigma^2{}_x}{2} t^{2H} + s^{2H} - (t-s)^{2H}. \tag{2.15}$$

in which is $\sigma^2{}_x = E(\boldsymbol{y}_t - \boldsymbol{y}_{t-1})^2 = E\boldsymbol{y}^2(1)$ is the variance of the $\boldsymbol{x}_t$ process of increments.

Consider the sampled version $\boldsymbol{y}_t$, $t \in \mathbb{Z}$, of a $\boldsymbol{y}_t$ H-sssi process, with unit sampling interval. There are many $\boldsymbol{y}_t$ H-sssi non-Gaussian processes. However, for value of $H \in (0,1)$ there is exactly just one $\boldsymbol{y}_t$ H-sssi Gaussian process, called Discrete-time Fractional Brownian Motion (DFBM) [9,97].

The Fractional Gaussian Noise (FGN), proposed by Mandelbrot and van Ness in 1968, corresponds to the DFBM process of increments. The FGN is a model widely used in LRD traffic simulations [6,31,93].

The DFBM and FGN models are non-parametric and are not used in traffic future values predictions.

[1] So, $E\boldsymbol{x}_t = E\boldsymbol{y}_t - \boldsymbol{y}_{t-1} = 0$

2.3.1 Exact second order self-similarity

Definition 2.3.1.1 (Exact second order self-similarity). *Consider the discrete time stationary process* $\mathbf{x}_t = \mathbf{y}_t - \mathbf{y}_{t-1}$. $\mathbf{x}_t$ *is an exact second order self-similar process with Hurst parameter* H $(1/2 < H < 1)$ *if its autocovariance exists and is given by [9]*

$$C_{\mathbf{x}}(\tau) = \frac{\sigma^2_{\mathbf{x}}}{2}|\tau + 1|^{2H} - 2|\tau|^{2H} + |\tau - 1|^{2H}, \quad \tau = \ldots, -1, 0, 1, \ldots \quad . \tag{2.16}$$

It can be shown that the autocovariance given by (2.16) satisfies [9]

$$\lim_{\tau \to \infty} \frac{C_x(\tau)}{\sigma_x^2 \tau^{2H-2} H(2H-1)} = 1, \tag{2.17}$$

i. e., $C_x(\tau)$ has a hyperbolic decay.

Second order self-similarity implies LRD when $1/2 < H < 1$.

Consider the aggregate process $X_t^{(M)}$ of an exact second order self-similar process x_t, at the M aggregation level. It can be shown that

$$C_x^{(M)}(\tau) = C_x(\tau), \qquad M = 2, 3, \ldots \tag{2.18}$$

Eq. (2.18) says that the second order statistics of the original process remains the same with a scale change, justifying the term "exact second order self similar".

Definition 2.3.1.2 (Asymptotic second order self-similarity). *A process* $\mathbf{x}_t$ *is asymptotically second order self-similar with Hurst parameter* H $(1/2 < H < 1)$ *if its autocovariance and its aggregate process autocovariance are related by [99]*

$$\lim_{M \to \infty} C_{\mathbf{x}}^{(M)}(\tau) = C_{\mathbf{x}}(\tau). \tag{2.19}$$

Tsybarov and Georganas have shown that (2.12) implies (2.19) [119]. Therefore, a LRD process is asymptotically second order self similar too.

2.3.2 Impulsiveness

Several phenomena exhibit impulsive behavior, as low frequency atmospheric noise, man-made noise, submarine acoustic noise, transmission lines transients, seismic activity [14, 79, 81, 100, 101], financial series [80], telephone circuits cluster errors [11], computer networks traffic [94]. In this context, the stable probability distribution [70] is a fundamental statistical modeling tool of impulsive signals [111]. Its use is justified by the generalized central limit theorem [88].

This theorem states that: if the limit of the sum of independent and identically distributed (i.i.d) random variables converges, then this limit can only be a stable distributed random variable.

Definition 2.3.2.1 (Heavy tail distribution). *A random variable* $\mathbf{x}$ *has a heavy tail distribution with index* α *if [109]*

$$P(\mathbf{x} \geq x) \sim cx^{-\alpha}L(x), \qquad x \to \infty, \tag{2.20}$$

for $c > 0$ *and* $0 < \alpha < 2$, *in which* $L(x)$ *is a positive function that varies slowly for large values of* x, *i. e.,* $\lim_{x\to\infty} L(bx)/L(x) = 1$ *for any positive* b.

Eq. (2.20) states that observations of a heavy tail distributed random variables may occur with non-negligible probabilities, with values very different from the mean (outliers). So, this kind of random variable has high variability.

A simple example of a heavy tail distribution is the Pareto I distribution [48], defined by means of its complementary distribution function (survival function):

$$\bar{F}(x) = P(\mathbf{x} \geq x) = \begin{cases} \left(\frac{x}{x_m}\right)^{-\alpha}, & x \geq x_m\,, \\ 1, & x < x_m\,, \end{cases} \tag{2.21}$$

Figures 2.10 and 2.11 show, respectively, the Pareto I probability density functions with $x_m = 1$ and the Pareto I probability distribution functions with $x_m = 1$.

The heavy distributions' p order statistics are finite if, and only if, $p \leq \alpha$. It is for this reason that such distributions have infinite variance. The mean is infinite if $\alpha < 1$. The heavy tail distributions are also known as infinite variance probability distributions.

An important member of the heavy tail distributions class is the *stable distribution*, discovered by Levy in the 1920 decade [70]. The stable distribution does not have an analytic expression[2]. It can be defined in terms of its characteristic function [109]:

$$\begin{aligned} \Phi_{\mathbf{x}}(w) = Ee^{jwx} &= \int_{-\infty}^{\infty} f_{\mathbf{x}}(x)e^{jwx}\,dx \\ &= \exp\{j\mu w - |\sigma w|^{\alpha}1 - j\eta\,\mathrm{sign}(w)\varphi(w,\alpha)\}. \end{aligned} \tag{2.22}$$

---

[2] The exceptions are the limit cases $\alpha = 1$ (Cauchy) and $\alpha = 2$ (Gaussian).

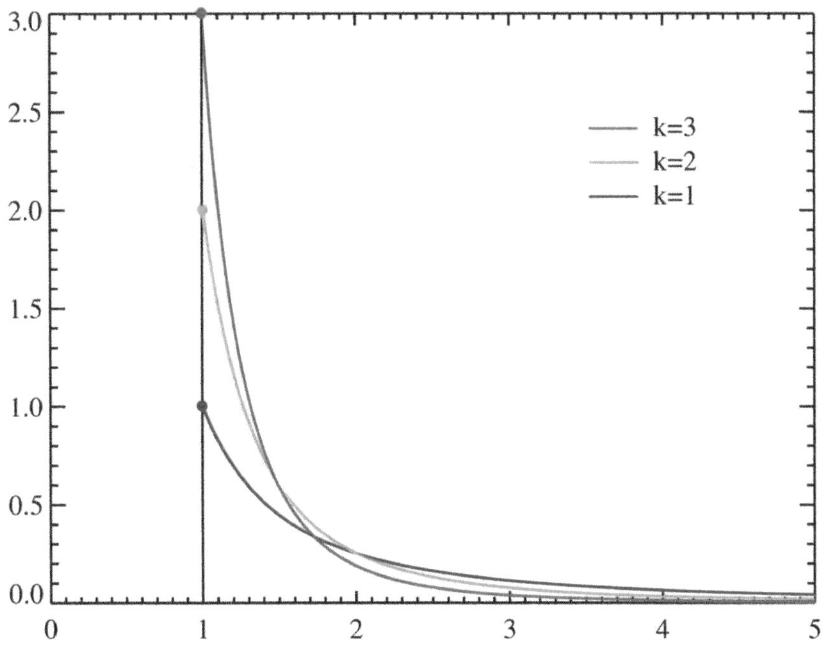

Figure 2.10: Pareto I probability density functions with $x_m = 1$.

In (2.22)

$$\varphi(w, \alpha) = \begin{cases} \tan(\alpha\pi/2) & \text{if } \alpha \neq 1 \\ -\frac{2}{\pi}\ln|w| & \text{if } \alpha = 1, \end{cases} \tag{2.23}$$

- sign(.) stands for the `sign` function,
- α $(0 < \alpha \leq 2)$ is the *characteristic exponent*,
- μ $(\mu \in \mathbb{R})$ is the *localization parameter*
- η $(-1 \leq \eta \leq 1)$ is the *asymmetry parameter* and
- $\sigma \geq 0$ is the *dispersion parameter* or *scale.*

Figure 2.12 shows a stable distribution realization example.

Figures 2.13 and 2.14 show a set of symmetric stable probability distributions and density functions examples.

Figures 2.15 and 2.16 show a set of asymmetric stable probability distributions and density functions examples.

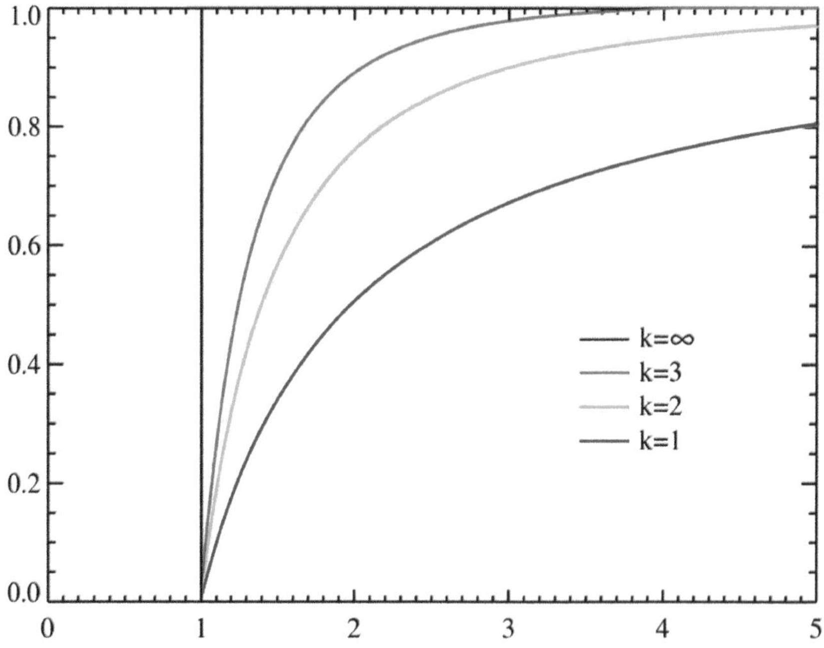

Figure 2.11: Pareto I probability distribution functions with $x_m = 1$.

It can be shown that, if $\boldsymbol{x} \sim S_\alpha(\sigma, \eta, \mu)$ with $0 < \alpha < 2$, then,

$$\begin{cases} \lim_{\lambda\to\infty} \frac{P(\boldsymbol{x}>\lambda)}{\lambda^{-\alpha}} = C_\alpha \frac{1+\eta}{2}\sigma^\alpha, \\ \lim_{\lambda\to\infty} \frac{P(\boldsymbol{x}<-\lambda)}{\lambda^{-\alpha}} = C_\alpha \frac{1-\eta}{2}\sigma^\alpha, \end{cases} \tag{2.24}$$

in which

$$C_\alpha = \left(\int_0^\infty x^{-\alpha} \sin x\right)^{-1} = \begin{cases} \frac{1-\alpha}{\Gamma(2-\alpha)\cos(\pi\alpha/2)} & \text{if } \alpha \neq 1 \\ 2/\pi & \text{if } \alpha = 1. \end{cases} \tag{2.25}$$

Therefore, (2.24) shows that the survival function of $\boldsymbol{x}$ decreases according to a power function for large values of λ.

Theorem 2.3.2.1 (Stability property). *A random variable* $\mathbf{x}$ *is stable if and only if for any independent random variables* $\mathbf{x}_1$ *and* $\mathbf{x}_2$ *that have the same distribution as* $\mathbf{x}$*, and for arbitrary constants* a_1, a_2*, there are constants* a *and* b *as*

$$a_1\mathbf{x}_1 + a_2\mathbf{x}_2 \stackrel{d}{=} a\mathbf{x} + b, \tag{2.26}$$

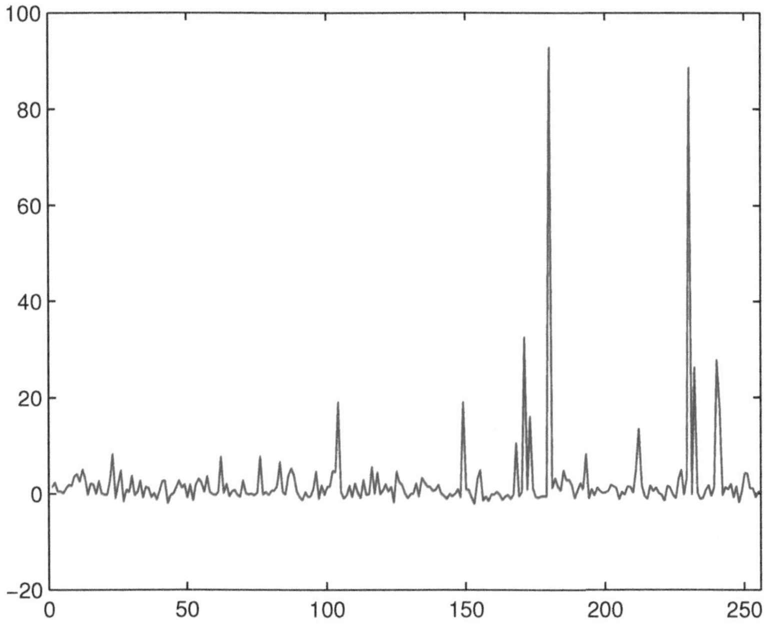

Figure 2.12: Realization of $S_\alpha(\sigma, \eta, \mu) = S_{1.2}(1, 1, 0)$ (256 samples).

Using the characteristic function of the stable distribution, it can be shown that: if $\boldsymbol{x}_1, \boldsymbol{x}_2, \ldots, \boldsymbol{x}_N$ are independent and follow stable distributions with the same (α, η), then all linear combinations $\sum_{j=1}^{N} a_j \boldsymbol{x}_j$ are stable with the same parameters α and η [88, 109, 111].

The central limit states that the normalized sum of (i.i.d) random variables with finite variance σ^2 and mean μ converges to a Gaussian distribution. Formally,

$$\frac{\bar{\boldsymbol{x}} - \mu}{\sigma/\sqrt{N}} \xrightarrow{d} \boldsymbol{x} \sim N(0, 1) \quad \text{for} \quad N \to \infty. \tag{2.27}$$

The relation (2.27) may be rewritten as

$$a_N(\boldsymbol{x}_1 + \boldsymbol{x}_2 + \ldots + \boldsymbol{x}_N) - b_N \xrightarrow{d} \boldsymbol{x} \sim N(0, 1) \quad \text{for} \quad \boldsymbol{x} \to \infty, \tag{2.28}$$

in which $a_N = 1/(\sigma\sqrt{N})$ and $b_N = \sqrt{N}\mu/\sigma$.

Theorem 2.3.2.2 (Generalized central limit theorem)**.** *Let* $\{\mathbf{x}_1, \mathbf{x}_2, \mathbf{x}_3 \ldots\}$ *be a sequence of i.i.d random variables. There are constants* $a_N > 0$, $b_N \in \mathbb{R}$

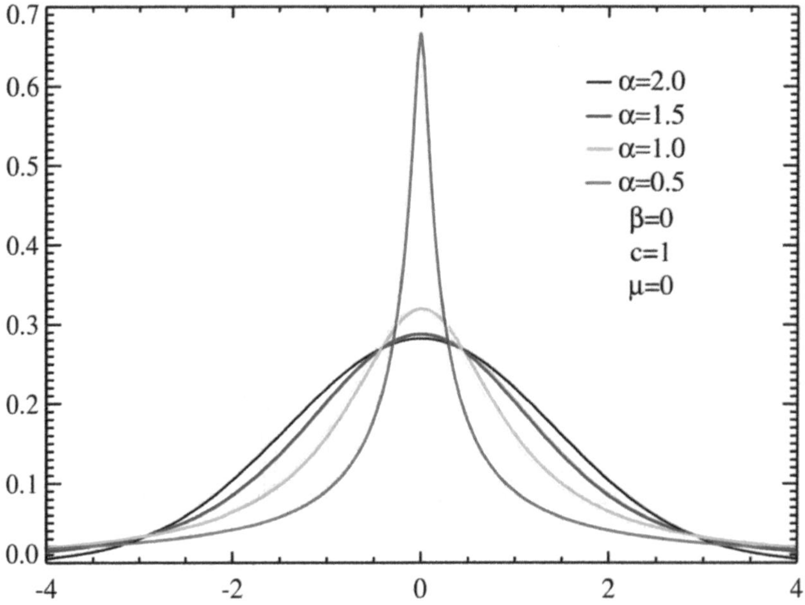

Figure 2.13: Symmetric stable distributions: density function.

and a random variable $\mathbf{x}$ *with*

$$a_N(\mathbf{x}_1 + \mathbf{x}_2 + \ldots + \mathbf{x}_N) - b_N \xrightarrow{d} \mathbf{x}$$

if and only if $\mathbf{x}$ *is* α-stable *with* $0 < \alpha \leq 2$.

Definition 2.3.2.2 (Impulsive stochastic process)**.** *A stochastic process* $\mathbf{x}_t$ *is impulsive if it has a heavy tail marginal probability distribution [109].*

Covariances (or correlations) cannot be defined in the stable random variables space. Remember, the stable random variable variance is infinite. Two kinds of measurements have been proposed [98, 111]:

- Co-variation;
- Co-difference;

Co-variation is not used in this book.

The co-difference of two jointly SαS random variables $\boldsymbol{x}_1$ and $\boldsymbol{x}_2$, $0 < \alpha \leq 2$, is given by

$$\gamma_{x_1,x_2} = (\sigma_{x_1})^\alpha + (\sigma_{x_2})^\alpha - (\sigma_{x_1-x_2})^\alpha, \tag{2.29}$$

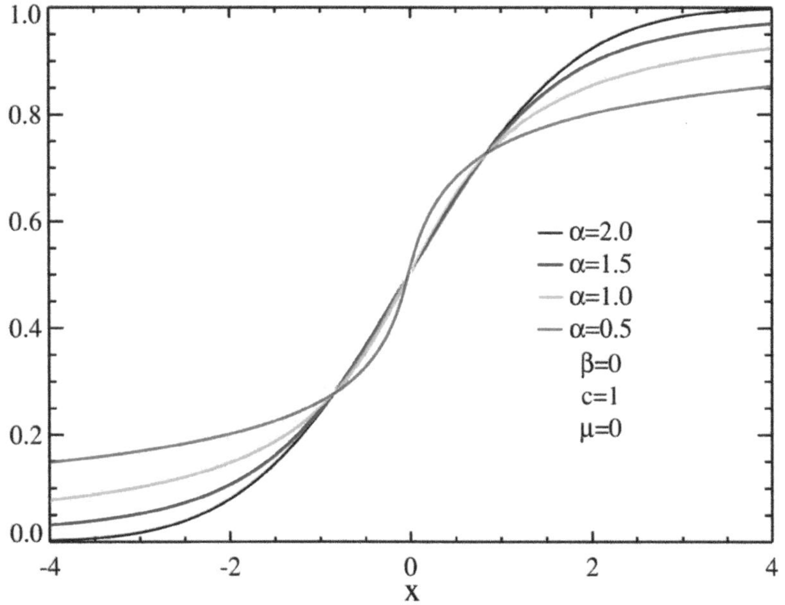

Figure 2.14: Symmetric stable distributions: probability distribution.

in which σ_x is the scale parameter of the SαS $\boldsymbol{x}$ random variable.

The co-difference is symmetric, i. e., $\gamma_{x_1,x_2} = \gamma_{x_2,x_1}$ and reduces to the covariance when $\alpha = 2$. If $\boldsymbol{x}_1$ e $\boldsymbol{x}_2$ are independent, then $\gamma_{x_1,x_2} = 0$.

Definition 2.3.2.3 (Generalized co-difference).

$$I(w_1, w_2; \mathbf{x}_1, \mathbf{x}_2) = -\ln Ee^{j(w_1\mathbf{x}_1 + w_2\mathbf{x}_2)} + \ln Ee^{jw_1\mathbf{x}_1} + \ln Ee^{jw_2\mathbf{x}_2}, \quad (w_1, w_2) \in \mathbb{R}^2 \tag{2.30}$$

If $\boldsymbol{x}_1$ and $\boldsymbol{x}_2$ are independent, then $I(w_1, w_2; \boldsymbol{x}_1, \boldsymbol{x}_2) = 0$. For jointly Gaussian random variables

$$I(w_1, w_2; \boldsymbol{x}_1, \boldsymbol{x}_2) = -w_1 w_2\, C(\boldsymbol{x}_1, \boldsymbol{x}_2),$$

in which $C(\boldsymbol{x}_1, \boldsymbol{x}_2)$ is the covariance between $\boldsymbol{x}_1$ and $\boldsymbol{x}_2$.

For stationary random processes we have

$$I(w_1, w_2; \tau) = I(w_1, w_2; \boldsymbol{x}_{t+\tau}, \boldsymbol{x}_t).$$

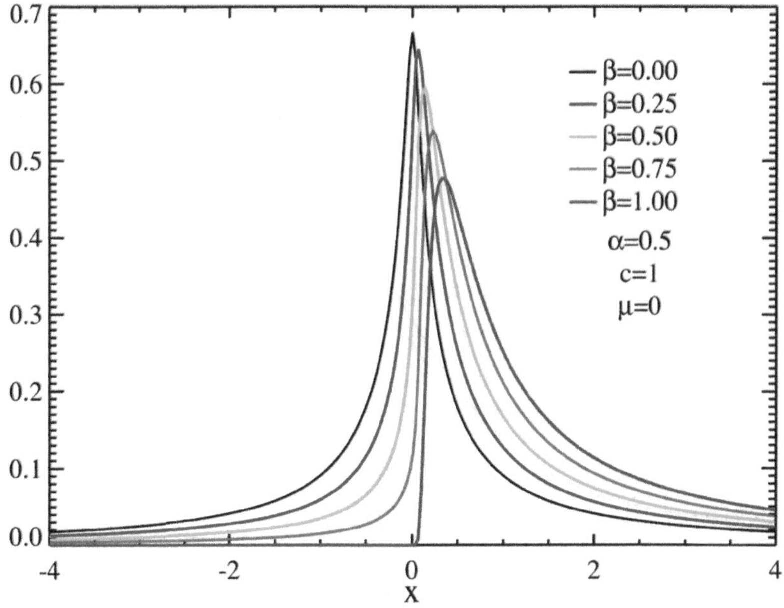

Figure 2.15: Asymmetric stable distributions: density function.

Definition 2.3.2.4 (Long memory in the generalized sense). *Let* $\{\mathbf{x}_t\}_{t\in\mathbb{R}}$ *be a stationary random process. We say that* $\mathbf{x}_t$ *has long memory - generalized sense if its generalized co-diiference* $I(w_1, w_2; \tau)|_{w_1=-w_2=1}$ *satisfies*

$$\lim_{\tau\to\infty} I(1,-1;\tau)/\tau^{-\beta} = L(\tau), \tag{2.31}$$

in which $L(\tau)$ *is a slow varying function for* $\tau \to \infty$ *and* $0 < \beta < 1$.

For LRD gaussian processes, the definition given above reduces to the classical definition of a LRD process.

2.4 Final remarks: why is the data networks traffic fractal?

Several authors [24, 36, 43, 69, 123, 124] have been stated that the Internet traffic self-similarity is caused by the packets size large variability of the individual sessions (FTP, HTTP, etc.) that make the aggregate traffic.

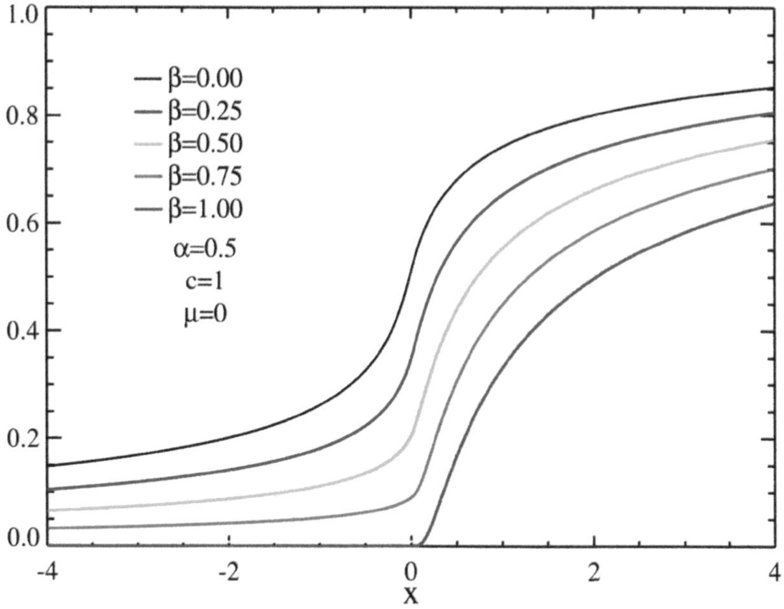

Figure 2.16: Asymmetric stable distributions: probability distribution.

Those papers argue that the IP traffic is self-similar because the individual sessions sizes that make the Internet traffic are generated by a heavy tail probability distribution.

On the other hand, the recent study by Gong *et al* [44] reexamines this question and claims that there is little evidence that the distribution's heavy tail has any impact on the algorithms' design and the Internet's infrastructure. Gong *et al* propose an on-off Markovian hierarchical model that explain the IP traffic's LRD and state that the multiple time scales involved in the traffic generation mechanism and transport protocols make the observation of LRD behavior unavoidable.

3

Modeling of long-range dependent teletraffic

This chapter introduces the concept of heterogenous and homogenous models. In addition, the difference behavioral and structural models are highlighted and their domains of application as well. The wavetet transform is introduced and used to generate the Multifractal Wavelet Model (MWM) model.

3.1 Classes of modeling

Traffic models may be classified as heterogenous or homogeneous. Heterogeneous models simulate the aggregate traffic (traffic generated by several users, protocols and applications) over a network link. Homogeneous models refer to a specific kind of traffic, as the MPEG video traffic [10, 22].

Heterogeneous models may be subdivided in two classes: behavioral [30, 73, 79, 93, 105] or structural [4, 124, 126].

- Behavioral models model the traffic statistics, as correlation, marginal distribution or even higher order statistics (third and fourth orders, for example) without taking into account the physical mechanism of traffic generation (i. e., behavioral models' parameters are not directly related to the communications network's parameters).
- Structural models are related to the packets generation mechanisms and their parameters may be mapped to network's parameters, as number of users and bandwidth.

FGN, Auto Regressive Fractionally Integrated and Moving Average (ARFIMA) and MWM are behavioral models of aggregate traffic. *On/Off* processes may be used as traffic structural models [99].

3.1.1 Non-parametric modeling

The FGN process, proposed by Mandelbrot and Van Ness in 1968 for modeling LRD hydrological series [79], is the first important long memory model that appears in the literature. If $\{\boldsymbol{x}_t\}_{t\in\mathbb{Z}}$ is a FGN, then $\boldsymbol{x}_t$ is a stationary process with autocovariance given by $C_x(\tau) = \frac{\sigma^2_x}{2}|\tau+1|^{2H} - 2|\tau|^{2H} + |\tau-1|^{2H}$, $\tau = \ldots, -1, 0, 1, \ldots$).

The FGN corresponds to the first difference of a continuous time stochastic process known as Fractional Brownian Motion (FBM) $\{\boldsymbol{B}_H(t) \quad 0 \leq t \leq \infty\}$ with Hurst parameter $0 < H < 1$, i. e., [97, p.279]

$$\boldsymbol{x}_t = \Delta \boldsymbol{B}_H(t) = \boldsymbol{B}_H(t+1) - \boldsymbol{B}_H(t), \quad t = 0, 1, 2, \ldots \quad . \tag{3.1}$$

Figure 3.1 illustrates some realizations of FBM processes for several Hurst parameters values.

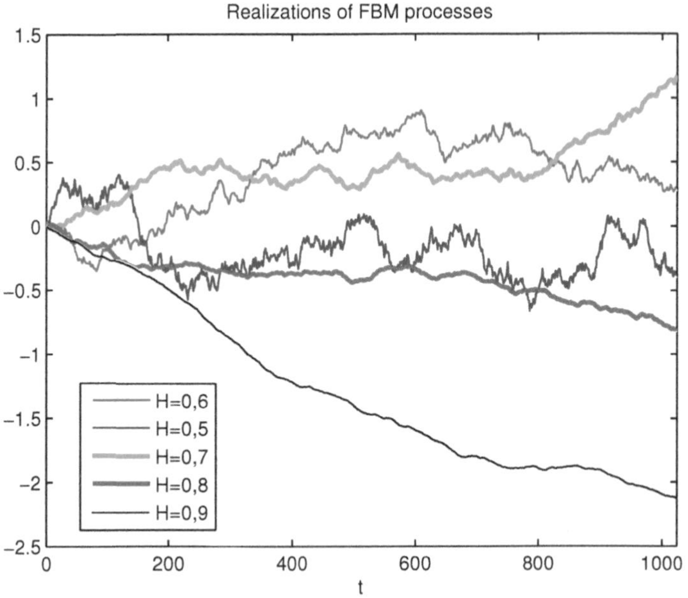

Figure 3.1: Realizations of FBM processes for several Hurst parameters values.

The FBM has a special name when $H = 1/2$: (*Brownian motion*) and it is designated by $\boldsymbol{B}_{1/2}(t)$. In this case, $\boldsymbol{x}_1, \boldsymbol{x}_2, \ldots$ are independent Gaussian random variables. We can create a discrete time FBM (DFBM), denoted by

$\boldsymbol{B}_t$, by means of the cumulative sum of the FGN $\{\boldsymbol{x}_t\}$ samples:

$$\boldsymbol{B}_t \equiv \boldsymbol{B}_H(t) = \sum_{u=0}^{t-1} \boldsymbol{x}_u, \quad t = 1, 2, \ldots \quad . \tag{3.2}$$

The DFBM SDF is given by [97, p. 280]

$$P_{\boldsymbol{B}_t}(f) = \sigma_x^2 C_H \sum_{j=-\infty}^{\infty} \frac{1}{|f + j|^{2H+1}}, \quad -\frac{1}{2} \leq f \leq \frac{1}{2}, \tag{3.3}$$

in which σ_x^2 is the power of a zero mean FGN, $C_H = \frac{\Gamma(2H+1)\sin(\pi H)}{2\pi^{2H+1}}$ and $0 < H < 1$.

According to (3.3), the DFBM SDF has a $|f|^{-\alpha}$, $0 < \alpha < 1$, singularity, in the origin, as

$$P_{\boldsymbol{B}_t} \propto |f|^{1-2H}, \quad f \to 0 . \tag{3.4}$$

The FGN and the DFBM are related by the transfer function

$$H(z) = \frac{X(z)}{B(z)} = 1 - z^{-1}, \tag{3.5}$$

in which $X(z)$ e $B(z)$ denote the z-transforms of x_t and B_t, respectively.

The frequency response associated to (3.5) is

$$H(f) = H(z)|_{z=e^{j2\pi f}} = 1 - e^{-j2\pi f} . \tag{3.6}$$

As the input/output relation in terms of the SDFs is [114, p. 351]

$$P_{\boldsymbol{x}}(f) = |H(f)|^2 P_{\boldsymbol{B}_t}(f) , \tag{3.7}$$

$|H(f)|^2$ is given by,

$$|H(f)|^2 = G(f) = 4\sin^2(\pi f) , \tag{3.8}$$

then the FGN's SDF is equal to

$$P_{\boldsymbol{x}}(f) = 4\sin^2(\pi f) P_{\boldsymbol{B}_t}(f) . \tag{3.9}$$

Equations (3.3) and (3.9) show that the FGN's SDF is characterized by only two parameters:σ_x^2 e H (responsible for the spectrum shape). The FGN is completely specified by its mean and by its SDF, as it is Gaussian.

It may be shown that (3.9) may be rewritten as [93]:

$$P_{\boldsymbol{x}}(f) = A(f, H)|2\pi f|^{-2H-1} + B(f, H) , \tag{3.10}$$

in which $A(f, H) = 2\sin(\pi H)\Gamma(2H+1)(1 - \cos(2\pi f))$ e $B(f, H) = \sum_{j=1}^{\infty}(2\pi j + 2\pi f)^{-2H-1} + (2\pi j - 2\pi f)^{-2H-1}$.

For small values of f, $P_{\boldsymbol{x}}(f) \propto |f|^{1-2H}$.

3.2 Wavelet transform

The Fourier Transform (FT) of a signal $x(t)$, if exists, is defined as

$$X(\nu) = \mathrm{TF}\{x(t)\} = \int_{-\infty}^{\infty} x(t)e^{-j2\pi\nu t}dt\,, \tag{3.11}$$

in which ν denotes the frequency in cycles/second [Hz].

Gabor [39] has shown that it is possible to represent the local spectral content of a signal $x(t)$ around an instant of time τ by the Windowed Fourier Transform (WFT).

$$X_T(\nu,\tau) = \int_{-\infty}^{\infty} x(t)g_T(t-\tau)e^{-j2\pi\nu t}dt\,, \tag{3.12}$$

in which $g_T(t)$ is a window of finite duration support T and ν denotes frequency.

The WFT is a bi-dimensional representation defined on the time-frequency plane or domain as it depends on the ν and τ parameters. The WFT would be equivalent to a kind of continuous "sheet music" description of $x(t)$.

According to the Heisenberg's uncertainty principle [59, p.52], a signal whose energy content is quite well localized in time has this energy quite spread out in the frequency domain. As the window of (3.12) has a fixed size T, we may conclude that the WFT is not good to analyze (or identify) behaviors of $x(t)$ occurring in time intervals much smaller or much larger than T, as, for example, transient phenomena of duration $\Delta t << T$ or cycles that exist in periods larger than T.

A *wavelet* $\psi_0(t)$ (sometimes also called mother wavelet), $t \in \mathbb{R}$, is a function that satisfies three conditions [41,97].

1. Its Fourier transform $\Psi(\nu)$, $-\infty < \nu < \infty$, is such that exists a finite constant C_ψ that obeys the *admissibility condition*

$$0 < C_\psi = \int_0^{\infty} \frac{|\Psi(\nu)|^2}{\nu} d\nu < \infty\,. \tag{3.13}$$

2. The integral of $\psi_0(t)$ is null:

$$\int_{-\infty}^{\infty} \psi_0(t)\,dt = 0\,. \tag{3.14}$$

3. Its energy is unitary:

$$\int_{-\infty}^{\infty} |\psi_0(t)|^2 \, dt = 1 \, . \tag{3.15}$$

Figure 3.2 shows four examples of wavelet functions: Haar, Daubechies, Coiflet and Symmlet. Figures 3.3, 3.4 and 3.5 illustrate, respectively the Meyer's wavelet, the Gaussian wavelet (related to the first derivative of a Gaussian PDF) and the "Mexican hat" wavelet (related to the second derivative of a Gaussian PDF).

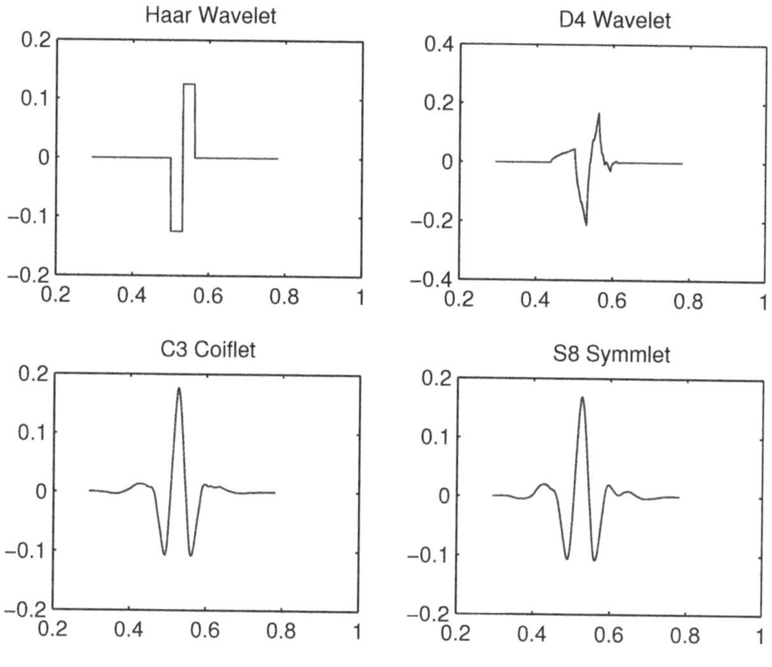

Figure 3.2: Four examples of wavelet functions.

The wavelet transform has been originally developed as an analysis and synthesis tool of continuous time energy signals [25, 26, 47, 75–77].

An energy signal $x(t)$, $t \in \mathbb{R}$ (t denotes time), obeys the constraint

$$\|x\|^2 = \langle x, x \rangle \equiv \int_{-\infty}^{\infty} |x(t)|^2 \, dt < \infty, \tag{3.16}$$

i. e., $x(t)$ that obeys the constraint (3.16) belongs to the squared summable functions space $L^2(\mathbb{R})$.

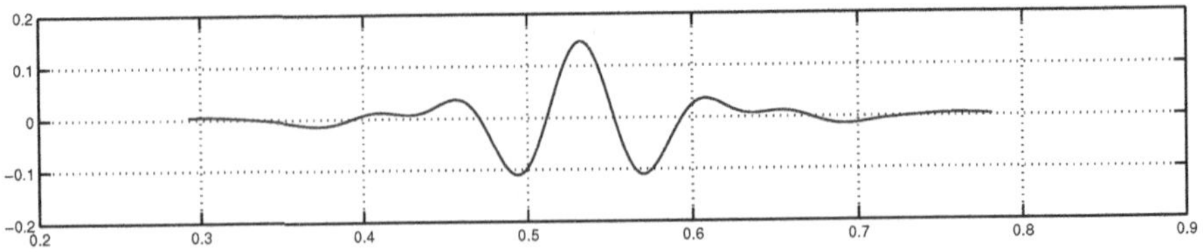

Figure 3.3: Meyer's wavelet.

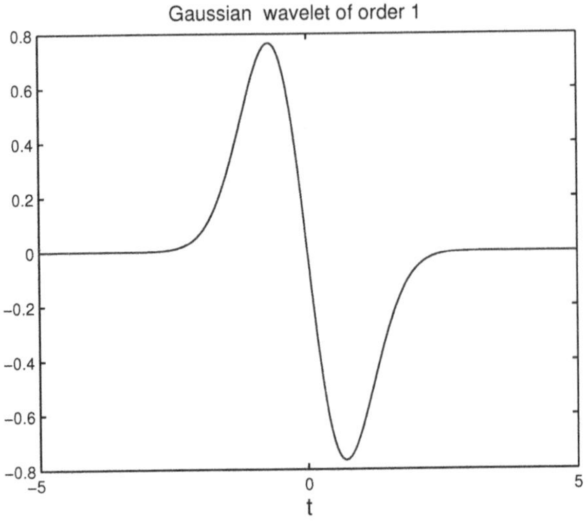

Figure 3.4: Gaussian wavelet (related to the first derivative of a Gaussian PDF).

Presently, the wavelet transform has also been used as an analysis tool of discrete time signals.

There are continuous time and discrete time wavelet decompositions designate by Continuous Wavelet Transform (CWT) and Discrete Wavelet Transform (DWT).

The CWT of a signal $x(t)$ consists of a set $C = \{W_{\psi}(s, \tau),\ s \in \mathbb{R}^{+},\ \tau \in \mathbb{R}\}$, in which

- τ is the time localization parameter,
- s represents scale and
- ψ denotes a wavelet function

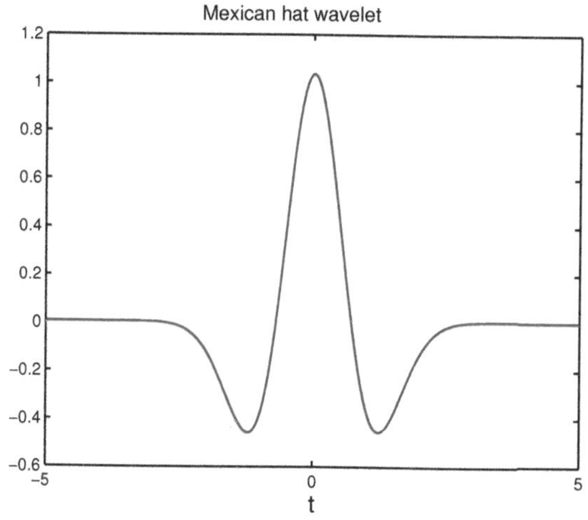

Figure 3.5: "Mexican hat" wavelet (related to the second derivative of a Gaussian PDF).

of wavelet coefficients on the continuous time-scale plane (also known as time-frequency plane)given by

$$W_\psi(s,\tau) = \langle \psi_{0_{(s,\tau)}}, x \rangle = \int_{-\infty}^{\infty} \frac{1}{\sqrt{s}} \psi_0^* \left(\frac{\lambda - \tau}{s} \right) x(\lambda) d\lambda \,. \tag{3.17}$$

$\psi_{0_{(s,\tau)}}(t) = s^{-1/2} \psi_0 \left(\frac{t-\tau}{s} \right)$ denotes a dilated and shifted version of the "mother" wavelet $\psi_0(t)$.

The factor $1/\sqrt{s}$ in (3.17) provides all functions of the class

$$\mathcal{W} = \left\{ \frac{1}{\sqrt{s}} \psi_0 \left(\frac{t-\tau}{s} \right) \in \mathbb{R} \right\} \tag{3.18}$$

have the same energy (norm).

The basic idea of the CWT defined by (3.17) is to correlate[1] a signal $x(t)$ with shifted (by τ) and dilated (by s) versions of a mother wavelet (that has a pass-band spectrum). The CWT is a two parameters function. So, it is a redundant transform, because it consists on mapping an one-dimension signal on the time-scale plane.

Differently from the WFT, where the reconstruction is made from the same family of functions as that used in the analysis, in the CWT the synthesis

[1] Measure the likelihood.

is made with functions $\tilde{\psi}_{s,\tau}$ that have to satisfy

$$\tilde{\psi}_{s,\tau}(t) = \frac{1}{C_\psi} \frac{1}{s^2} \psi_{s,\tau}(t) \,. \tag{3.19}$$

So, $x(t)$ is completely recovered by the Inverse Continuous Wavelet Transform (ICWT):

$$x(t) = \frac{1}{C_\psi} \int_0^\infty \left[\int_{-\infty}^{\infty} W_\psi(s,\tau) \frac{1}{\sqrt{s}} \psi\left(\frac{t-\tau}{s}\right) d\tau \right] \frac{ds}{s^2} \,. \tag{3.20}$$

The fundamental difference between the CWT and the WFT consists of the fact that the functions $\psi_{s,\tau}$ undergo dilations and compressions [59]. The analysis on refined scales of time (small values of s) requires "fast" $\psi_{s,\tau}$ functions, i. e., of a small support, while the analysis on aggregate scales of time (large values of s) requires "slower" $\psi_{s,\tau}$ functions, i. e., of a wider support. As already mentioned, the internal product defined by (3.17) is a likelihood measure between the wavelet $\psi\left(\frac{t-\tau}{s}\right)$ and the signal $x(t)$ on a certain instant of time τ and on a determined scale s. For a fixed τ, large values of s correspond to a low-frequency analysis, while small values of s are associated to a high-frequency analysis. Therefore, the wavelet transform has a *variable time resolution* (i. e., the capacity of analyzing a signal from close - "*zoom in*" - or from far - "*zoom out*"), being adequate to analyze phenomena that occur in different time scales.

Figure 3.6 provides an example of a CWT.

3.2.1 Multiresolution analysis and the discrete wavelet transform

There are two kinds of DWT:

- the DWT for discrete time signals and
- the DWT for continuous time signals.

The DWT may be formulated for discrete time signals (as it is done, for example, by Percival and Walden [97]) without establishing any explicit connection with the CWT. On the other hand, we should not understand the term "discrete" of the DWT for continuous time signals as meaning that this transform is defined over a discrete time signal. But only that the coefficients produced by this transform belong to a subset $D = \{w_{j,k} = W_\psi(2^j, 2^j k),\ j \in \mathbb{Z},\ k \in \mathbb{Z}\}$ of the set C [120], [41, p.105].

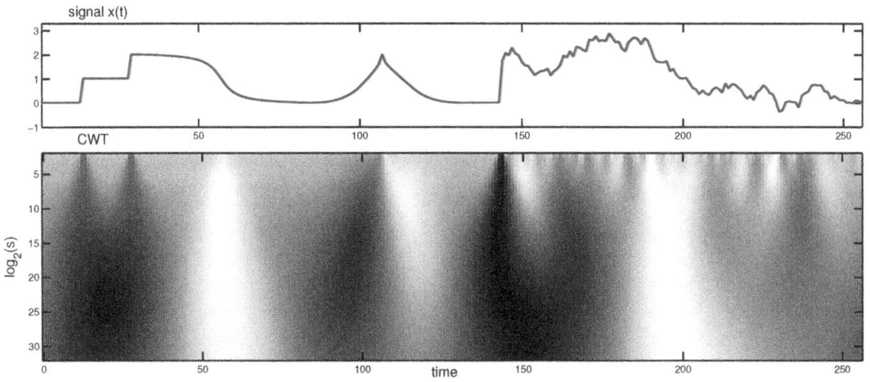

Figure 3.6: The image on the bottom part of the figure is the CWT $W_\psi(s, \tau)$ of the signal on the top part.

In fact, the DWT coefficients for continuous time signals can also be directly obtained by means of the integral

$$w_{j,k} = \left\langle \psi_{0_{(2^j, 2^j k)}}, x \right\rangle = \int_{-\infty}^{\infty} 2^{-j/2} \psi_0^*(2^{-j}\lambda - k) x(\lambda)\, d\lambda\,, \qquad (3.21)$$

in which the indices j and k are called scale and localization, respectively, does not involve any discrete time signal, but the continuous time signal $x(t)$.

Equation (3.21) shows that the continuous time DWT corresponds to a critically sampled version of the CWT defined by (3.17) in the dyadic scales $s = 2^j$, $j = \ldots, -1, 0, 1, 2, \ldots$, in which the instants of time in the dyadic scale $s = 2^j$ are separated by multiples of 2^j. The function ψ_0 of (3.21) must be defined from a MultiResolution Analysis (MRA) of the signal $x(t)$ [25, 74, 97]. Observe that the continuous time MRA theory is similar to that of discrete time.

Although the teletraffic signals are discrete time, we decided to present the continuous time MRA version because the Hurst parameter estimator based on wavelets proposed by Abry and Veitch [1] is based on the spectral analysis of a "fictitious" process $\{\tilde{\boldsymbol{x}}_t,\ t \in \mathbb{R}\}$ that is associated to the discrete time process $\{\boldsymbol{x}_n,\ n \in \mathbb{Z}\}$ [120].

Figure 3.7 shows the critical sampling of the time-scale plane by means of the CWT parameters ($s = 2^j$ e $\tau = 2^j k$) discretization.

A MRA is, by definition, a sequence of closed subspaces $\{V_j\}_{j\in\mathbb{Z}}$ de $L^2(\mathbb{R})$ such that [97, p.462], [25]:

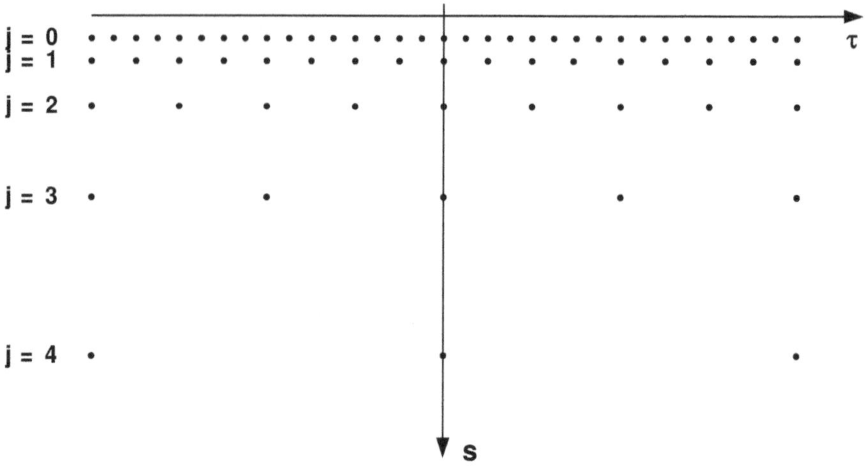

Figure 3.7: Critical sampling of the time-scale plane by means of the CWT parameters ($s = 2^j$ e $\tau = 2^j k$) discretization.

1. $\ldots V_2 \subset V_1 \subset V_0 \subset V_{-1} \subset V_{-2} \subset \ldots$;
2. $\bigcap_{j \in Z} V_j = \{\}$;
3. $\bigcup_{j \in Z} V_j = L^2(\mathbb{R})$;
4. $x(t) \in V_j \Leftrightarrow x(2^j t) \in V_0,\ j > 0$ (in which t denotes time and $x(t)$ is an energy signal);
5. There is a function $\phi_j(t) = 2^{-j/2}\phi_0(2^{-j}t)$ in V_j, called *scale function*, such that the set $\{\phi_{j,k},\ k \in \mathbb{Z}\}$ is an orthonormal basis of V_j, with $\phi_{j,k}(t) = 2^{-j/2}\phi_0(2^{-j}t - k)\ \forall j, k \in \mathbb{Z}$.

The subspace $V_j$is known as the *approximation space* associated to the time scale $s_j = 2^j$ (assuming that V_0 is the approximation space with unit scale).

If the $x(t)$ projection on V_j is represented by the scale coefficients

$$u_{j,k} = \langle \phi_{j,k}, x \rangle = \int_{-\infty}^{\infty} 2^{-j/2}\phi_0^*(2^{-j}t - k)x(t)\,dt, \qquad (3.22)$$

then the properties 1 and 3 assure that $\lim\limits_{j \to -\infty} \sum_k \phi_{j,k}(t)u_{j,k} = x(t),\ \forall\, x \in L^2(\mathbb{R})$.

Property 4 implies that the subspace V_j is a scaled version of subspace V_0 (multiresolution).

The orthonormal basis mentioned in property 5 is obtained by time shifts of the low-pass function ϕ_j.

Consider the successive approximations sequence (also known in the literature as *wavelet smooths* [97]) of $x(t)$

$$\mathcal{S}_j(t) = \sum_k \phi_{j,k}(t) u_{j,k} \quad j = \ldots, -1, 0, 1, \ldots . \tag{3.23}$$

As $V_{j+1} \subset V_j$, we have $\mathcal{S}_{j+1}(t)$ is a coarser approximation of $x(t)$ than $\mathcal{S}_j(t)$.

This fact illustrates the MRA's fundamental idea, that consists in examining the *loss of information* when one goes from $\mathcal{S}_j(t)$ to $\mathcal{S}_{j+1}(t)$:

$$\mathcal{S}_j(t) = \mathcal{S}_{j+1}(t) + \Delta x_{j+1}(t). \tag{3.24}$$

$\Delta x_{j+1}(t)$ (called *detail* of $x_j(t)$) belongs to the subspace W_{j+1}, named detail space [97] that is associated to the fluctuations (or variations) of the signal in the more refined time scale $s_j = 2^j$ and that corresponds to the orthogonal complement of V_{j+1} in V_j[2].

The MRA shows that the detail signals $\Delta x_{j+1}(t) = \mathcal{D}_{j+1}(t)$ may be directly obtained by successive projections of the original signal $x(t)$ on wavelet subspaces W_j.

Besides, the MRA theory shows that exists a function $\psi_0(t)$, called "mother wavelet" , that is obtained from $\phi_0(t)$, in which $\psi_{j,k}(t) = 2^{-j/2}\phi_0(2^{-j}t - k)$ $k \in \mathbb{Z}$ is an orthonormal basis of W_j.

The detail $\mathcal{D}_{j+1}(t)$ is obtained by the equation

$$\mathcal{D}_{j+1}(t) = \sum_k \psi_{j+1,k}(t) \left\langle \psi_{j+1,k}(t), x(t) \right\rangle . \tag{3.25}$$

The internal product $\left\langle \psi_{j+1,k}(t), x(t) \right\rangle = w_{j+1,k}$ denotes the wavelet coefficient associated to scale $j+1$ and discrete time k and $\{\psi_{j+1,k}(t)\}$ is a family of wavelet functions that generates the subspace W_{j+1}, orthogonal to subspace V_{j+1} ($W_{j+1} \perp V_{j+1}$), i. e.,

$$\left\langle \psi_{j+1,n}, \phi_{j+1,p} \right\rangle = 0 \, , \forall n, p. \tag{3.26}$$

Therefore, the detail signal $\mathcal{D}_{j+1}(t)$ belongs to the complementary subspace W_{j+1} de V_j, because

$$V_j = V_{j+1} \oplus W_{j+1}. \tag{3.27}$$

---

[2] Besides, W_{j+1} is contained in the subspace V_j.

That is, V_j is given by the direct addition of V_{j+1} and W_{j+1}, and this means that any element in V_j may be determined from the addition of two orthogonal elements belonging to V_{j+1} and W_{j+1}. Iterating (3.27), we have

$$V_j = W_{j+1} \oplus W_{j+2} \oplus \dots \quad . \tag{3.28}$$

Eq. (3.28) says that the approximation $\mathcal{S}_j(t)$ is given by

$$\mathcal{S}_j(t) = \sum_{i=j+1}^{\infty} \sum_{k} w_{i,k} \psi_{i,k}(t) \, . \tag{3.29}$$

The MRA of a continuous time signal $x(t)$ is initiated by determining the coefficients[3] $u_0(k) = \langle \phi_{0,k}(t), x(t) \rangle$, in which $k = 0, 1, \dots, N-1$, that are associated to the projection of $x(t)$ on the approximation subspace V_0.

Following, the sequence $\{u_0(k)\}$ is decomposed by filtering and sub-sampling by a factor of 2 (*downsampling*) in two sequences: $\{u_1(k)\}$ and $\{w_1(k)\}$, each one with $N/2$ points. This filtering and sub-sampling process is repeated several times, producing the sequences

$$\{\{u_0(k)\}_N, \{u_1(k)\}_{\frac{N}{2}}, \{u_2(k)\}_{\frac{N}{4}}, \dots, \{u_j(k)\}_{\frac{N}{2^j}}, \dots, \{u_J(k)\}_{\frac{N}{2^J}}\} \tag{3.30}$$

and

$$\{\{w_1(k)\}_{\frac{N}{2}}, \{w_2(k)\}_{\frac{N}{4}}, \dots, \{w_j(k)\}_{\frac{N}{2^j}}, \{w_J(k)\}_{\frac{N}{2^J}}\} \, . \tag{3.31}$$

The literature calls the set of coefficients [1, 120]

$$\left\{\{w_1(k)\}_{\frac{N}{2}}, \{w_2(k)\}_{\frac{N}{4}}, \dots, \{w_J(k)\}_{\frac{N}{2^J}}, \{u_J(k)\}_{\frac{N}{2^J}}\right\} \tag{3.32}$$

as the DWT of the $x(t)$ signal.

Fig. 3.8 illustrates a 3-levels DWT (decomposition in scales $j = 1, 2, 3$) associated to 1024 samples of the discrete time $x(k) = \sin(3k) + \sin(0.3k) + \sin(0.03k)$, that corresponds to the superposition of 3 sinusoids in frequencies $f_1 \approx 0.004775$, $f_2 \approx 0.04775$ and $f_3 \approx 0.4775$. Fig. 3.9 shows the SDF of this signal. The graph concatenates the sequences of the scale coefficients $\{u_3(k)\}_{128}$ and of the wavelet coefficients $\{w_3(k)\}_{128}$, $\{w_2(k)\}_{256}$ e $\{w_1(k)\}_{512}$ from left to right, i. e., the first 128 points correspond to the sequence $\{u_3(k)\}_{128}$; then follow the 128 points of the sequence $\{w_3(k)\}_{128}$, the 256 points of the sequence $\{w_2(k)\}_{256}$ and 512 points of the sequence $\{w_1(k)\}_{512}$.

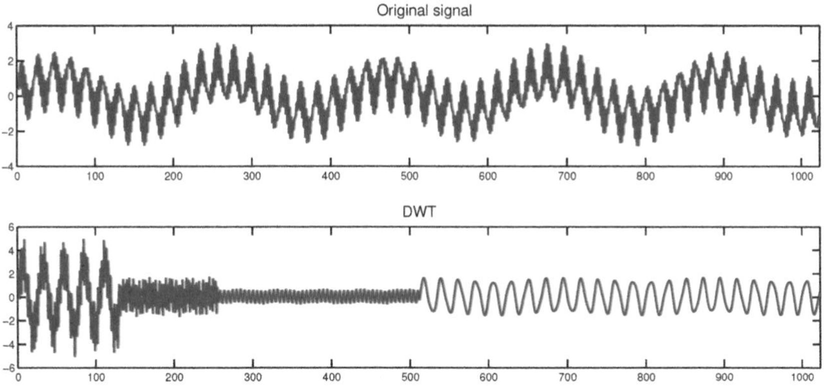

Figure 3.8: An illustration of the 3-levels DWT of the discrete time signal $x(k) = \sin(3k) + \sin(0.3k) + \sin(0.03k)$.

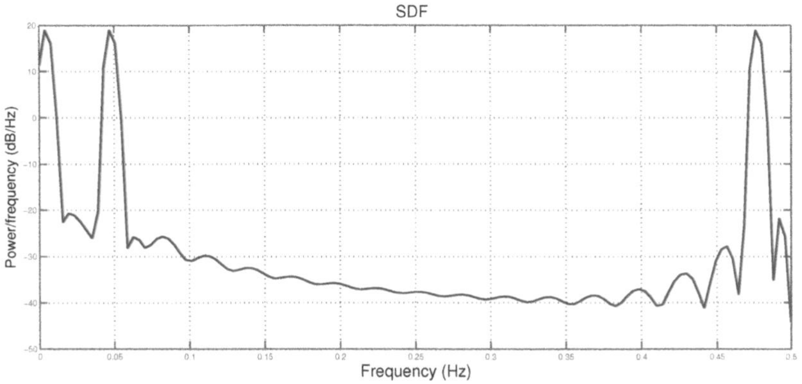

Figure 3.9: SDF of the signal $x(k) = \sin(3k) + \sin(0.3k) + \sin(0.03k)$.

The reconstruction of $x(t)$ is implemented by filtering and oversampling by a factor of 2 (*upsampling*) of the sequences (3.30) and (3.31), obtaining an approximation of $x(t)$ in the subspace V_0

$$\mathcal{S}_0(t) = \mathcal{S}_J(t) + \mathcal{D}_1(t) + \mathcal{D}_2(t) + \cdots + \mathcal{D}_J(t) \tag{3.33}$$

[3] The sequence $u_0(k)$ is obtained sampling the filter's output whose impulse response is $\phi^*(-t)$ (matched filter with a function $\phi_0(t) = \phi(t)$) at instants $k = 0, 1, 2, \ldots$, i. e., $u_0(k) = x(t) \star \phi^*(-t)$ for $k = 0, 1, 2, \ldots$, in which $\star$ denotes convolution.

or

$$x(t) \approx \sum_{k} u(J,k)\phi_{J,k}(t) + \sum_{j=1}^{J}\sum_{k} w_{j,k}\psi_{j,k}(t)\,. \qquad (3.34)$$

Eq. (3.34) defines the Inverse Discrete Wavelet Transform (IDWT) and Figure 3.10 shows the synthesis of the signal $x(k) = \sin(3k) + \sin(0.3k) + \sin(0.03k)$ in terms of the sum $\mathcal{S}_3(t) + \mathcal{D}_1(t) + \mathcal{D}_2(t) + \mathcal{D}_3(t)$.

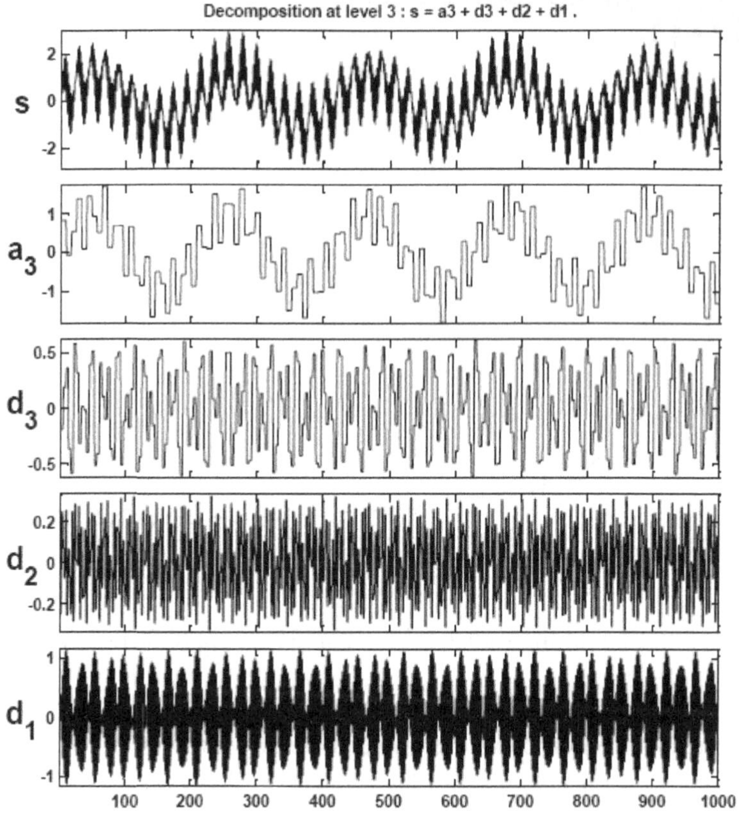

Figure 3.10: Synthesis of the signal $x(k) = \sin(3k) + \sin(0.3k) + \sin(0.03k)$ in terms of the sum $\mathcal{S}_3(t) + \mathcal{D}_1(t) + \mathcal{D}_2(t) + \mathcal{D}_3(t)$.

We say that the function $\phi_0(t) = \phi(t)$ determines a MRA of $x(t)$ according to (3.33), if it obeys the following conditions:

1. intra-scale orthonormality (property 5)

$$\langle \phi(t-m), \phi(t-n) \rangle = \delta_{m,n} , \tag{3.35}$$

   in which $\delta_{m,n}$ is the Kronecker's delta ($\delta_{m,n} = 1$ if $m = n$, $\delta_{m,n} = 0$ for $m \neq n$). Eq. (3.35) imposes an orthonormality condition at scale $j = 0$.
2. unit mean

$$\int_{-\infty}^{\infty} \phi(t)\, dt = 1 . \tag{3.36}$$

3.

$$\frac{1}{\sqrt{2}}\phi(\frac{t}{2}) = \sum_n g_n \phi(t-n) , \tag{3.37}$$

   as several $\phi(t-k)$ fit in $\phi(\frac{t}{2})$ (is a consequence of property (1) of the MRA).

Equation 3.37 may be rewritten as

$$\phi(t) = \sum_n \sqrt{2} g_n \phi(2t-n) , \tag{3.38}$$

known as *Dilation Equation.*

Equations 3.37 and 3.38 may be written, respectively, in the frequency domain as

$$\sqrt{2}\Phi(2\nu) = G(\nu)\Phi(\nu) , \tag{3.39}$$

and

$$\Phi(\nu) = \frac{1}{\sqrt{2}} G(\nu)\Phi(\frac{\nu}{2}) , \tag{3.40}$$

in which $\Phi(\nu)$ is the Fourier transform of $\phi(t)$ and $G(\nu) = \sum_n g_n e^{-j2\pi \nu n}$, known as *scale filter* (low-pass), represents a periodic filter in ν.

As the subspace W_{j+1} is orthogonal to V_{j+1} and is in V_j, we have

$$\frac{1}{\sqrt{2}}\psi(\frac{t}{2}) = \sum_n h_n \phi(t-n) , \tag{3.41}$$

or

$$\psi(t) = \sum_n \sqrt{2} h_n \phi(2t-n) , \tag{3.42}$$

that is the *Wavelet Equation.*

Applying the Fourier transform to (3.41) and (3.42) we get, respectively,

$$\sqrt{(2)}\Psi(2\nu) = H(\nu)\Phi(\nu) , \tag{3.43}$$

and

$$\Psi(\nu) = \frac{1}{\sqrt{2}} H(\nu)\Phi(\frac{\nu}{2}) . \tag{3.44}$$

in which $H(\nu)$ is the *wavelet filter* (high-pass).

Rewriting (3.26) in terms of the frequency domain and using (3.39) and (3.43) results the *orthogonality condition*

$$\int_{-\infty}^{\infty} G(\nu)H^*(\nu)|\Phi(\nu)|^2 \, d\nu = 0 , \tag{3.45}$$

that the filter H has to obey so the family $\{\psi_{1,k}(t)\}$ is orthogonal to the family $\{\phi_{1,k}(t)\}$.

We may show that the condition [59, p.150], [97, p.75]

$$h_n = (-1)^n g_{L-1-n} , \quad \leftrightarrow \quad H(z) = -z^{-L+1} G(-z^{-1}) , \tag{3.46}$$

in which L denotes the length of a FIR filter g_n, is sufficient to (3.45) to hold.

We say that g_n e h_n are Quadrature Mirrored Filters (QMF) when they are related by (3.46).

Figure 3.11 shows the QMF filters frequency response (graphs on the upper part) *vs brickwall*-type filters frequency response.

According to (3.38), the MRA departs from a definition (from several possible) of the scale function $\phi(t)$, that is related to the scale filter g_n by (3.37). Eq. (3.46) says that the choice of a Finite Impulse Response (FIR)-type filter $\{g_n\}$ implies a $\{h_n\}$ that is also FIR. At last, the wavelet function is determined by (3.41).

The scale $\phi(t)$ and wavelet $\psi(t)$ functions associated to the FIR filters $\{g_n\}$ e $\{h_n\}$ have compact support, thus offering the *time resolution* functionality. The simplest scale function that satisfies (3.35) is the characteristic function of the interval $I = 0, 1)$, that corresponds to the Haar's scale function:

$$\phi^{(H)}(t) = \chi_{0,1)}(t) = \begin{cases} 1 & \text{se } 0 \le t < 1 \\ 0 & \text{otherwise.} \end{cases} \tag{3.47}$$

In this case (Haar MRA), the associated Haar scale filter is given by

$$g_n = \{\ldots, 0, g_0 = 1/\sqrt{2}, g_1 = 1/\sqrt{2}, 0, \ldots\} , \tag{3.48}$$

the Haar wavelet filter by

$$h_n = \{\ldots, 0, h_0 = g_1 = 1/\sqrt{2}, h_1 = -g_0 = -1/\sqrt{2}, 0, \ldots\} \tag{3.49}$$

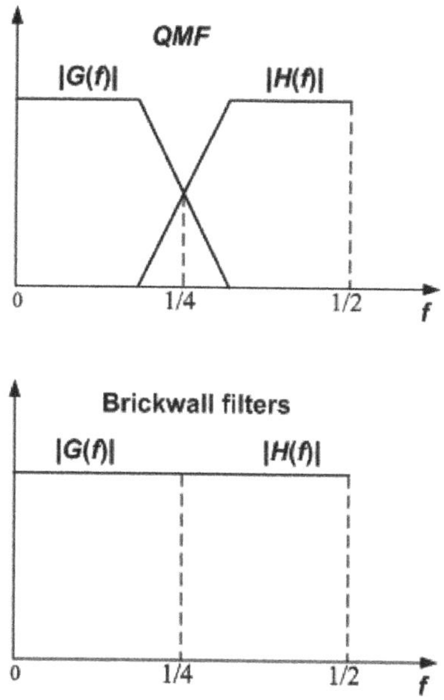

Figure 3.11: QMF filters frequency response (graphs on the upper part) *vs* *brickwall*-type filters frequency response.

and the Haar wavelet function by

$$\psi^{(H)}(t) = \chi_{0,1/2)}(t) - \chi_{1/2,1)}(t)\,. \tag{3.50}$$

Figure 3.12 shows the Daubechies' scale and wavelet functions with $N = 2, 3, 4$ vanishing moments

$$\int_{-\infty}^{\infty} t^m \psi(t)\, dt = 0, \quad m = 0, 1, \ldots, N-1\,. \tag{3.51}$$

Ingrid Daubechies [26] was the first one to propose a method for building sequences of transfer functions $\{G^{(N)}(z)\}_{N=1,2,3,\ldots}$ and $\{H^{(N)}(z)\}_{N=1,2,3,\ldots}$, in which $G^{(N)}(z)$ is associated to the low-pass FIR filter $g_n^{(N)}$ and $H^{(N)}(z)$ to the high-pass filter $h_n^{(N)}$. The corresponding scale and wavelet functions have support in $0, 2N-1$. The first member of the sequence is the Haar system $\phi^{(1)} = \phi^{(H)}$, $\psi^{(1)} = \psi^{(H)}$. The Daubechies' filters are generalizations of the Haar system for $N \geq 2$ [59].

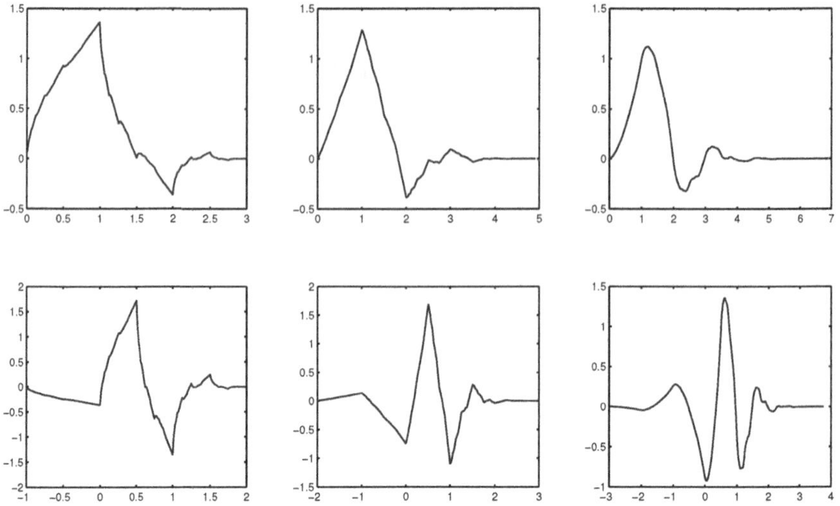

Figure 3.12: Daubechies' wavelets with $N = 2, 3, 4$ *vanishing moments*, bottom row from left to right, respectively. The corresponding scale functions are in the upper part.

We can demonstrate that [77]:

$$u_j(n) = \sum_k g(k - 2n)u_{j-1}(k) \tag{3.52}$$

and that

$$w_j(n) = \sum_k h(k - 2n)u_{j-1}(k)\,. \tag{3.53}$$

According to (3.52) and (3.53), we can obtain the coefficients $u_j(n)$ and $w_j(n)$ from the scale coefficients $u_{j-1}(m)$ by means of decimation operation of the sequence $\{u_{j-1}(m)\}$ by a factor of 2. The decimation consists in cascading a low-pass filter $g(-m)$ (with a transfer function $\bar{G}(z) = G(1/z)$ and frequency response $G^*(f)$) or a high-pass $h(-m)$ (with transfer function $\bar{H}(z) = H(1/z)$ and frequency response $H^*(f)$) with a compressor (or decimator) by a factor of 2. Decimate a signal by a factor D is the same as to reduce its sampling rate by D times.

The MRA is implemented by a low-pass and high-pass analysis filter banks $G^*(f)$ and $H^*(f)$ adequately positioned for separating the scale and wavelet coefficients sequences. Later, it is possible to rebuild the original signal using dual QMF reconstruction filter banks, low-pass $G(f)$ and high-pass $H(f)$.

It is important to emphasize that the pyramid algorithm's complexity is $O(N)$ (assuming we want to evaluate the DWT of N samples), while the direct evaluation of the DWT (that involves matrices multiplication) is $O(N^2)$ [97].

Figure 3.13 shows the QMF analysis filter banks $G^*(f)$ (low-pass) and $H^*(f)$ (high-pass) with decimation (*downsampling*) by a factor of 2. Figure 3.14 shows the QMF reconstruction filter banks with interpolation (*upsampling*) by a factor of 2. Observe that are used dual low-pass and high-pass filters, $G(f)$ and $H(f)$.

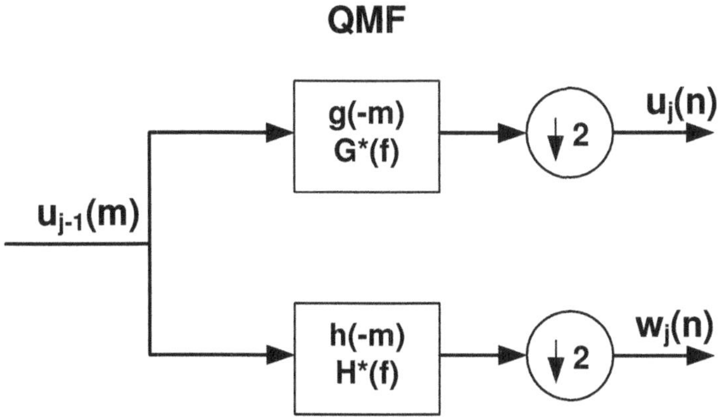

Figure 3.13: QMF analysis filter banks $G^*(f)$ (low-pass) and $H^*(f)$ (high-pass) with decimation (*downsampling*) by a factor of 2.

Figure 3.15 presents the flow diagram that shows the initial projection of a signal $x(t)$ on V_0 followed by the decomposition in W_1, W_2 and V_2.

Figure 3.16 shows the flow diagram that illustrates the approximate synthesis of $x(t)$ from W_1, W_2 and V_2.

Figure 3.17 presents a block diagram that shows that the DWT works as a sub-bands codification scheme. The spectrum $U_0(f)$ of the signal $u_0(n)$ is subdivided in three frequency bands (that cover two octaves): $0 \le f < 1/8$, $1/8 \le f < 1/4$ and $1/4 \le f \le 1/2$.

3.3 Model MWM

The MWM uses the Haar's MRA and is based on a multiplicative binomial cascade in the wavelet domain, that ensures that the simulated series are pos-

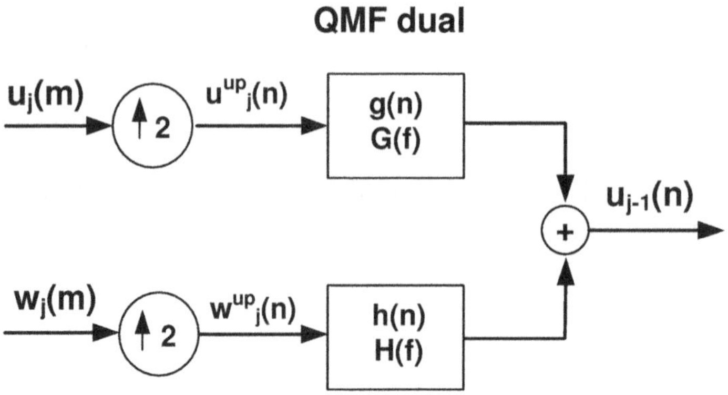

Figure 3.14: QMF reconstruction filter banks with interpolation (*upsampling*) by a factor of 2. Observe that are used dual low-pass and high-pass filters, $G(f)$ and $H(f)$.

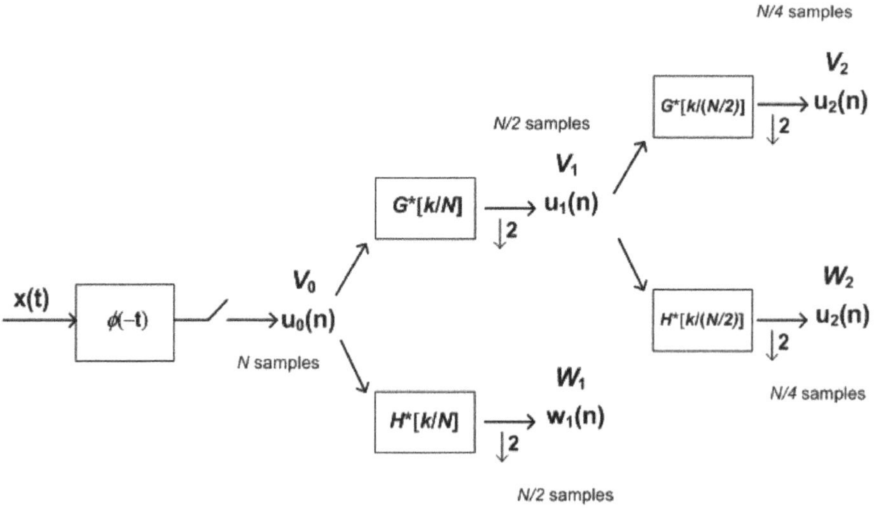

Figure 3.15: Flow diagram that shows the initial projection of a signal $x(t)$ on V_0 followed by the decomposition in W_1, W_2 and V_2.

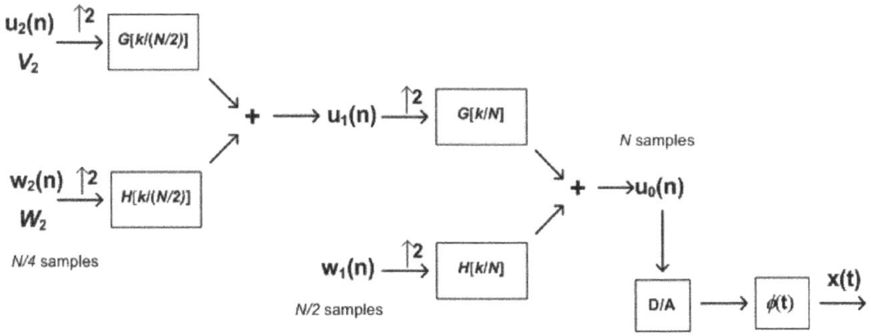

Figure 3.16: Flow diagram that illustrates the approximate synthesis of $x(t)$ from W_1, W_2 and V_2.

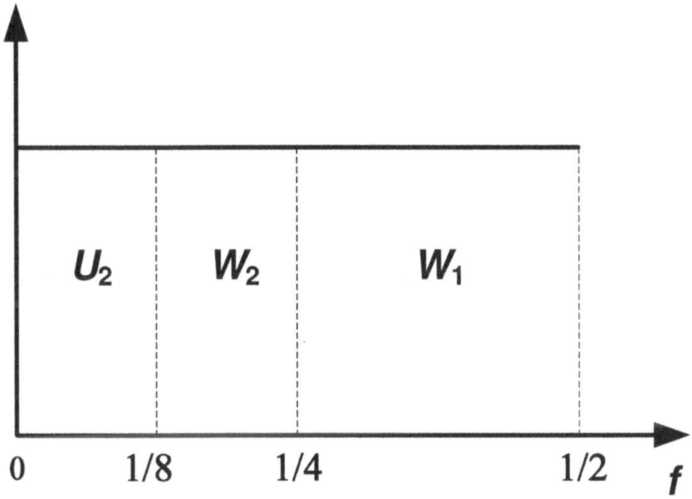

Figure 3.17: Block diagram that shows that the DWT works as a sub-bands codification scheme. The spectrum $U_0(f)$ of the signal $u_0(n)$ is subdivided in three frequency bands (that cover two octaves): $0 \leq f < 1/8$, $1/8 \leq f < 1/4$ and $1/4 \leq f \leq 1/2$.

itive [105]. The binomial cascade is a random binomial tree whose root is the coefficient $u_{J-1,0}$ (the MWM considers that $\frac{N}{2^{J-1}} = 1$, where N is the number of samples)

$$x_t = x_{0,k} = u_{J-1,0}\phi_{J-1,0}(t) + \sum_{j=1}^{J-1}\sum_{k} w_{j,k}\psi_{j,k}(t), \tag{3.54}$$

in which $\phi_{J-1,0}(t)$ denotes a Haar's scale function in the slowest scale (order $J-1$) and $w_{j,k}$ are the wavelet coefficients.

The Haar's scale and wavelet coefficients may be recursively computed by the following set of synthesis equations,

$$u_{j-1,2k} = 2^{-1/2}(u_{j,k} + w_{j,k}) \tag{3.55}$$

$$u_{j-1,2k+1} = 2^{-1/2}(u_{j,k} - w_{j,k}). \tag{3.56}$$

So, strictly positive signals may be modeled if $u_{j,k} \geq 0$ and

$$|w_{j,k}| \leq u_{j,k} \,. \tag{3.57}$$

It is possible to choose a statistical model for $w_{j,k}$ that incorporates the condition (3.57).

The MWM specifies a multiplicative model,

$$w_{j,k} = M_{j,k}u_{j,k} \,, \tag{3.58}$$

in which the multiplier $M_{j,k}$ may be modeled as a random variable with symmetric β distribution with shape parameter p_j, i. e., $M_j \sim \beta(p_j, p_j)$.

In this case, the MWM is known as β-MWM and we assume that the $M_{j,k}$'s are mutually independent and independent from $u_{j,k}$[4].

The variance of M_j is given by

$$\mathrm{Var}M_j = \frac{1}{2p_j + 1} \,. \tag{3.59}$$

In this way, equations (3.55) and (3.56) may be rewritten as

$$u_{j-1,2k} = \left(\frac{1 + M_{j,k}}{\sqrt{2}}\right) u_{j,k} \tag{3.60}$$

$$u_{j-1,2k+1} = \left(\frac{1 - M_{j,k}}{\sqrt{2}}\right) u_{j,k}. \tag{3.61}$$

[4] Riedi et al have also investigated other distributions for the multipliers.

These equations show that the MWM is, in fact, a binomial cascade.

The MWM may approximate the SDF of a training sequence by modeling the variance decay of the wavelet coefficients

$$\eta_j = \frac{\mathrm{Var} w_{j,k}}{\mathrm{Var} w_{j-1,k}} = \frac{2p_{(j-1)} + 1}{p_{(j)} + 1}, \tag{3.62}$$

that leads to

$$p_{(j-1)} = \frac{\eta_j}{2}(p_{(j)} + 1) - 1/2 \tag{3.63}$$

and

$$p_{(j)} = \frac{2p_{(j-1)} + 1}{\eta_j} - 1\,. \tag{3.64}$$

The MWM assumes that $u_{J-1,0}$ (the "root" scale coefficient) is approximately Gaussian. We can show that $p_{(j)}$ converges to

$$p_{-\infty} = \lim_{j \to -\infty} p_{(j)} = \frac{2^{\alpha} - 1}{2 - 2^{\alpha}}\,. \tag{3.65}$$

Table 3.1 shows the asymptotic values of the shape parameter p of the β multipliers as function of α (or H).

Table 3.1: Asymptotic values of the shape parameter p of the β multipliers as function of α (or H).

α	0.1	0.2	0.5	0.8
p	0.077	0.175	0.707	2.86
H	0.55	0.6	0.75	0.9

The MWM model has multifractal properties and the marginal probability density is lognormal.

3.4 Parametric modeling

3.4.1 ARFIMA model

The modeling of a (linear) times series x_t consists of estimating a transfer function (or model) $H(B)$ such that

$$\mathbf{x}_t = H(B)\mathbf{w}_t, \tag{3.66}$$

in which $\mathbf{w}_t$ is the innovation at instant t.

In practice, the modeling is based on the estimation of the inverse function $G(B) = H(B)^{-1}$, because we expect that filtering x_t by $G(B)$ produces a series of residuals w_t of the WN type.

Granger and Joyeux [45], and Hosking [53] introduced, in a independent way, the class of models ARFIMA[5] that has the following properties:

1. long memory explicit modeling;
2. flexibility for modeling the series' autocorrelation structure for small and large lags;
3. enable to simulate LRD series from the model.

Consider the equation

$$\Delta^d \boldsymbol{x}_t = \boldsymbol{w}_t, \tag{3.67}$$

in which d is a fractionary exponent, $0 < d < 1/2$.

Observe that

$$\Delta^d = (1 - B)^d = \sum_{k=0}^{\infty} \binom{d}{k} (-1)^k B^k, \tag{3.68}$$

with binomial coefficients[6]

$$\binom{d}{k} = \frac{\Gamma(d+1)}{\Gamma(k+1)\Gamma(d-k+1)}, \tag{3.69}$$

results the *fractionary difference filter*

$$\Delta^d = 1 - dB + \frac{1}{2}d(d-1)B^2 - \frac{1}{3}d(d-1)(d-2)B^3 + \ldots, \tag{3.70}$$

that is defined for any real $d > -1$.

According to (3.70), the model (3.67) is of the kind AR(∞).

Equation (3.67) defines the *fractionary integrated process* (also called FD(d) model or *fractionary WN* that is an extension of the integrated model ARIMA($0, d, 0$), $d \in \mathbb{Z}_+$ [83, 97]. The FD process is able to model the $1/f^\alpha$ singularity at the origin of a LRD series' spectrum. The FD is stationary and LRD when $0 < d < 1/2$; it is stationary and SRD when $-1/2 < d < 0$; it is non-stationary[7] when $|d| > 1/2$.

---

[5] The ARFIMA model can also be designated as FARIMA. Both terms are used interchangeably in this book.

[6] The Gamma function extends the factorial function for real and complex numbers: $d = \Gamma(d+1)$.

[7] In this case, $\boldsymbol{x}_t$ has infinite variance [45].

In practice, we observe that the SACF's decay for *small values of the lag* of some real teletraffic series is well modeled by SRD processes, i. e., have significative values that exponentially decay for small values of the lag, which is not well modeled by the FD(d) process [28, 30, 31, 73, 93, 105].

This does not mean that this kind of teletraffic series is not asymptotically LRD; it only says that the SRD characteristic may manifest itself by the existence of local SDF peaks (besides the singularity at the origin of the spectrum that is due to the long memory), as illustrated by Figure 3.18.

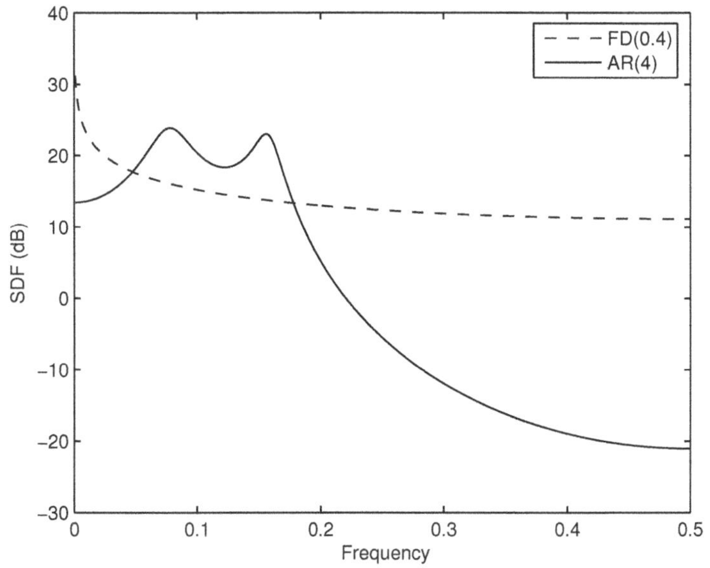

Figure 3.18: SDFs for equal power AR(4) and FD(0,4).

It is for this reason that there is a need to introduce the ARFIMA(p, d, q) class of models (more flexible than the FD class)

$$\phi(B)\Delta^d \mathbf{x}_t = \theta(B)\mathbf{w}_t, \tag{3.71}$$

in which $-1/2 < d < 1/2$, and

- $\phi(B)$ is the p order auto-regressive operator;
- $\theta(B)$ is the q order moving average operator;
- $\mathbf{w}_t$ is a Gaussian WN.

The model (3.71) is *LRD, stationary and invertible* when $0 < d < 1/2$ and if the poles and zeroes of $\theta(z)/\phi(z)$ are inside the unit radius circle. The parameter d models the high order autocorrelation structure (in which the decay is slow, of the hyperbolic type). On the other hand, the parameters of the polynomials $\phi(B)$ and $\theta(B)$ are responsible for modeling the low order lags autocorrelation (fast decay of the exponential type).

Equation (3.71) may be rewritten as an AR(∞)

$$\frac{\phi(B)}{\theta(B)}\Delta^d \boldsymbol{x}_t = \boldsymbol{w}_t. \tag{3.72}$$

The SDF of an ARFIMA(p, d, q) model is given by [28, 83]

$$P_x(f) = \frac{{\sigma^2}_w|1 - e^{-j2\pi f}|^{-2d}|1 - \theta_1 e^{-j2\pi f} - \ldots - \theta_q e^{-jq2\pi f}|^2}{|1 - \phi_1 e^{-j2\pi f} - \ldots - \phi_p e^{-jp2\pi f}|^2}, \tag{3.73}$$

in which ${\sigma^2}_w$ is the power of $\boldsymbol{w}_t$ and $w = 2\pi f$ is the normalized angular frequency ($-\pi \le w \le \pi$).

Equation (3.73) may be simplified for the case ARFIMA($0, d, 0$):

$$P_x(f) = |1 - e^{-j2\pi f}|^{-2d}\,{\sigma^2}_w = 2(1 - \cos(2\pi f))^{-d}\,{\sigma^2}_w \tag{3.74}$$

or

$$P_x(f) = 2\sin(\pi f/2)^{-2d}\,{\sigma^2}_w. \tag{3.75}$$

As $\sin w \approx w$ for w close to zero, then (3.75) reduces to

$$P_x(f) = (\pi f)^{-2d}\,{\sigma^2}_w. \tag{3.76}$$

3.4.2 ARFIMA models prediction - optimum estimation

Consider the three basic formats of the ARIMA(p, d, q) model at instant $t + h$:

(a) ARMA($p + d, q$)

$$\boldsymbol{x}_{t+h} = \sum_{k=1}^{p+d} \varphi_k \boldsymbol{x}_{t+h-k} + \boldsymbol{w}_{t+h} - \sum_{k=1}^{q} \theta_k \boldsymbol{w}_{t+h-k}; \tag{3.77}$$

(b) AR(∞)

$$\boldsymbol{x}_{t+h} = \sum_{k=1}^{\infty} g_k \boldsymbol{x}_{t+h-k} + \boldsymbol{w}_{t+h}; \tag{3.78}$$

(c) MA(∞)[8]

$$\boldsymbol{x}_{t+h} = \sum_{k=0}^{\infty} \psi_k \boldsymbol{w}_{t+h-k}. \tag{3.79}$$

Let

$$\hat{x}_{t+h} = \psi_h^* w_t + \psi_{h+1}^* w_{t-1} + \psi_{h+2}^* w_{t-2} + \ldots \tag{3.80}$$

be the Minimum Mean Squared Error (MMSE) estimated sequence. Then the coefficients ψ_{h+k}^*, $k = 0, 1, 2, \ldots$, may be determined minimizing the prevision's Mean Square Error (MSE)

$$E(\mathbf{e}_{t+h})^2 = E(\boldsymbol{x}_{t+h} - \hat{\boldsymbol{x}}_{t+h})^2 = E\left[\left(\sum_{k=0}^{\infty} \psi_k \boldsymbol{w}_{t+h-k} - \sum_{k=0}^{\infty} \psi_{h+k}^* \boldsymbol{w}_{t+h-k}\right)^2\right]. \tag{3.81}$$

As

$$\sum_{k=0}^{\infty} \psi_k \boldsymbol{w}_{t+h-k} = \sum_{k=-h}^{\infty} \psi_{h+k} \boldsymbol{w}_{t-k},$$

we have that $\mathbf{e}_{t+h}$ is given by

$$\mathbf{e}_{t+h} = \psi_0 \boldsymbol{w}_{t+h} + \psi_1 \boldsymbol{w}_{t+h-1} + \ldots + \psi_{h-1} \boldsymbol{w}_{t+1} - \sum_{k=0}^{\infty} (\psi_{h+k} - \psi_{h+k}^*) \boldsymbol{w}_{t-k}. \tag{3.82}$$

In this way,

$$E(\mathbf{e}_{t+h})^2 = (1 + \psi_1^2 + \ldots + \psi_{h-1}^2) \sigma^2{}_w + \sum_{k=0}^{\infty} (\psi_{h+k} - \psi_{h+k}^*)^2 \sigma^2{}_w, \tag{3.83}$$

in which $\psi_0 = 1$, because the innovations $\boldsymbol{w}_t$ are non-correlated.

It follows that $\psi_{h+k}^* = \psi_{h+k}$ minimizes (3.83). Therefore, the optimum prediction according the MMSE criterium is given by

$$\hat{x}_{t+h} = \psi_h w_t + \psi_{h+1} w_{t-1} + \psi_{h+2} w_{t-2} + \ldots = \sum_{k=0}^{\infty} \psi_{h+k} w_{t-k} \tag{3.84}$$

and the minimum prediction error by

$$\mathbf{e}_{t+h} = \boldsymbol{w}_{t+h} + \psi_1 \boldsymbol{w}_{t+h-1} + \ldots + \psi_{h-1} \boldsymbol{w}_{t+1} \tag{3.85}$$

---

[8] The sequence ψ_t in (3.79) denotes the ARIMA model's impulse response.

has a variance

$$V_h = (1 + \psi_1^2 + \ldots + \psi_{h-1}^2)\sigma_w^2, \tag{3.86}$$

given that

$$E\mathbf{e}_{t+h}|\mathcal{F}_t^{(\infty)} = 0, \tag{3.87}$$

in which $\mathcal{F}_t^{(\infty)} = \{x_t, x_{t-1}, \ldots\}$ denotes the set of all past observations of the series.

Note that

$$\boldsymbol{x}_{t+h} = \hat{x}_{t+h} + \mathbf{e}_{t+h}, \quad h \geq 1. \tag{3.88}$$

Observe that the conditional expectation of $\boldsymbol{x}_{t+h}$ given the series past observations

$$E\boldsymbol{x}_{t+h}|\mathcal{F}_t^{(\infty)} = \hat{x}_{t+h}, \tag{3.89}$$

is equal to the MMSE prediction (see (3.87)) (this does not happen by chance).

In fact, we can demonstrate that the optimum predictor according to the MMSE criterium [110, 114]:

a) is the conditional expectation $E\boldsymbol{x}_{t+h}|\mathcal{F}_t^{(\infty)}$ and
b) that this is a linear predictor when the innovations are Gaussian.

3.4.3 Forms of prediction

Taking the conditional expectation in (3.77), we get the prediction by means of the differences equation

$$\begin{aligned}\hat{x}_{t+h} &= \varphi_1 E\boldsymbol{x}_{t+h-1}|\mathcal{F}_t^{(\infty)} + \ldots + \varphi_{p+d} E\boldsymbol{x}_{t+h-p-d}|\mathcal{F}_t^{(\infty)} \\ &\quad + E\boldsymbol{w}_{t+h}|\mathcal{F}_t^{(\infty)} - \theta_1 E\boldsymbol{w}_{t+h-1}|\mathcal{F}_t^{(\infty)} - \ldots - \theta_q E\boldsymbol{w}_{t+h-q}|\mathcal{F}_t^{(\infty)},\end{aligned} \tag{3.90}$$

for $h \geq 1$.

Observe that

$$\begin{aligned} E\boldsymbol{x}_{t+k}|\mathcal{F}_t^{(\infty)} &= \hat{x}_{t+k}, \quad k > 0, \\ E\boldsymbol{x}_{t+k}|\mathcal{F}_t^{(\infty)} &= x_{t+k}, \quad k \leq 0, \\ E\boldsymbol{w}_{t+k}|\mathcal{F}_t^{(\infty)} &= 0, \quad k > 0, \\ E\boldsymbol{w}_{t+k}|\mathcal{F}_t^{(\infty)} &= w_{t+k}, \quad k \leq 0. \end{aligned} \tag{3.91}$$

Note that [83, p.227]

(a) $\hat{x}_{t+h}$ depends on $\hat{x}_{t+h-1}, \hat{x}_{t+h-2}, \ldots$, that are evaluated in a recursive way;

(b) in practice, we only know a finite number of past observations, i. e., $\mathcal{F}_t = \{x_t, x_{t-1}, \ldots, x_1\}$. Therefore,

$$E\boldsymbol{x}_{t+k}|\mathcal{F}_t^{(\infty)} \approx E\boldsymbol{x}_{t+k}|\mathcal{F}_t;$$

(c) the predictions for an AR(p) are exact, because we can show that

$$E\boldsymbol{x}_{t+k}|x_t, x_{t-1}, \ldots = E\boldsymbol{x}_{t+k}|x_t, \ldots, x_{t+1-p}$$

Taking the conditional expectation in (3.78) we obtain the prediction in the AR(∞) format

$$\hat{x}_{t+h} = \sum_{k=1}^{\infty} g_k E\boldsymbol{x}_{t+h-k}|\mathcal{F}_t^{(\infty)} + E\boldsymbol{w}_{t+h}|\mathcal{F}_t^{(\infty)}. \tag{3.92}$$

As $E\boldsymbol{w}_{t+h}|\mathcal{F}_t^{(\infty)} = 0$, we can rewrite (3.92) in the format

$$\hat{x}_{t+h} = g_1\hat{x}_{t+h-1} + g_2\hat{x}_{t+h-2} + \ldots + g_h x_t + g_{h+1}x_{t-1} + \ldots \quad . \tag{3.93}$$

3.4.4 Confidence interval

The ARFIMA model (3.71) assumes that the sequence of innovations $\boldsymbol{w}_t$ is a null mean WGN, i. e., $\boldsymbol{w}_t \sim \mathcal{N}(0, \sigma_w^2)$. It then follows that the conditional distribution of $\boldsymbol{x}_{t+h}$ given $\mathcal{F}_t^{(\infty)}$ is of the type $\mathcal{N}(\hat{x}_{t+h}, V_h)$ and that

$$Z = \frac{x_{t+h} - \hat{x}_{t+h}}{\sqrt{V_h}} \sim \mathcal{N}(0, 1). \tag{3.94}$$

The expression of the confidence interval for $\boldsymbol{x}_{t+h}$, at the confidence level $(1 - \beta)$, is given by

$$\hat{x}_{t+h} - z_{\beta/2}\sqrt{V_h} \leq \boldsymbol{x}_{t+h} \leq \hat{x}_{t+h} + z_{\beta/2}\sqrt{V_h}. \tag{3.95}$$

As in practice the value of σ_w^2 is unknown, we use the estimate

$$\hat{V}_h = (1 + \psi_1^2 + \ldots + \psi_{h-1}^2)\hat{\sigma}_w^2, \tag{3.96}$$

obtained in the model estimation phase.

At last, we obtain the final expression of the confidence interval for $\boldsymbol{x}_{t+h}$

$$\hat{x}_{t+h} - z_\beta\hat{\sigma}_w\left[1 + \sum_{k=1}^{h-1}\psi_k^2\right]^{1/2} \leq \boldsymbol{x}_{t+h} \leq \hat{x}_{t+h} + z_\beta\hat{\sigma}_w\left[1 + \sum_{k=1}^{h-1}\psi_k^2\right]^{1/2}. \tag{3.97}$$

3.4.5 ARFIMA prediction

Consider the stationary and invertible ARFIMA(p, d, q) model, $-0,5 < d < 0,5$, given by (3.72). We can rewrite the process in the format AR(∞)

$$\sum_{k=0}^{\infty} g_k \boldsymbol{x}_{t-k} = \boldsymbol{w}_t, \tag{3.98}$$

in which $g_0 = 1$ and

$$\sum_{k=0}^{\infty} g_k B^k = \phi(B)\theta^{-1}(B)(1-B)^d = \pi(B). \tag{3.99}$$

Then, we can predict a future value of $\boldsymbol{x}_t$ using (3.99) and (3.93) [83, p.474]. The prediction error variance is given by (3.86). Note that the polynomial $\pi(B)$ has infinite order (as $|d| < 1/2$). As in practice we have a series with N observations, only the first L terms of $\pi(B)$ are used, with $L < N$.

3.5 Long memory statistical tests

3.5.1 R/S statistics

Consider a time series x_t, $t = 1, 2, \ldots, N$. Hurst proposed the long memory test [54]

$$Q_N = \frac{1}{\hat{s}_N}\left[\max_{1\le k\le N}\sum_{j=1}^{k}(x_j - \bar{x}) - \min_{1\le k\le N}\sum_{j=1}^{k}(x_j - \bar{x})\right], \tag{3.100}$$

in which $\hat{s}_N = \sqrt{\hat{C}_0}$, known as *Range Over Standard Deviation (R/S)* statistics or *rescaled adjusted range* [9, 128].

Hurst observed that the R/S log-log plot (for the Nile river's yearly minimal levels time series) *versus* N spread along a straight line with slope greater than 1/2, i. e., that log (R/S) versus N presented a kind of CN^H behavior (Hurst effect), in which C is a constant and $1/2 < H < 1$ denotes the Hurst parameter.

This empirical discovery contradicts the expected behavior of Markovian processes (that are SRD), in which R/S must have an asymptotical behavior of $CN^{1/2}$.

The statistics $N^{-1/2}Q_N$ converges to a well defined random variable (for $N \to \infty$) when $\boldsymbol{x}_t$ is a WGN process [128, p.262], [9, p.82]. That is the reason why the R/S log-log plot *versus* N presents a $CN^{1/2}$ asymptotical behavior . On the other hand, it is the $N^{-H}Q_N$ statistics that converges to a well defined random variable when $\boldsymbol{x}_t$ is LRD [9].

Afterwards, Lo [72] has shown that (3.100) is not robust when SRD is also present in the series and developed an extended version of (3.100)

$$Q_T = \frac{1}{\hat{\sigma}_{NW}}\left[\max_{1\leq k\leq N}\sum_{j=1}^{k}(x_j - \bar{x}) - \min_{1\leq k\leq N}\sum_{j=1}^{k}(x_j - \bar{x})\right], \tag{3.101}$$

in which $\hat{\sigma}_{NW}$ denotes the square root of the long run variance Newey-West estimate of a process $\boldsymbol{x}_t$ (stationary and ergodic) [128, p.85], [49].

The long run variance is defined as

$$lrv(\boldsymbol{x}_t) = \sum_{\tau=-\infty}^{\infty} C_\tau. \tag{3.102}$$

As $C_{-\tau} = C_\tau$, (3.102) may be rewritten as

$$lrv(\boldsymbol{x}_t) = C_0 + 2\sum_{\tau=1}^{\infty} C_\tau. \tag{3.103}$$

The Newey-West estimator for (3.102) is given by

$$\widehat{lrv}_{NW}(x_t) = \hat{C}_0 + 2\sum_{\tau=1}^{T} w_{\tau,T}\hat{C}_\tau, \tag{3.104}$$

in which $w_{\tau,T}$ are coefficients (whose sum is equal to one) and a truncate parameter that satisfies $T = O(N^{1/3})$.

3.5.2 GPH test

Geweke and Porter-Hudak [42] proposed a long memory test based on the SDF of the ARFIMA$(0, d, 0)$ process given by

$$P_x(f) = 4\sin^2(\pi f)^{-d}\sigma^2{}_w, \tag{3.105}$$

in which $\sigma^2{}_w$ denotes the power of the $\boldsymbol{w}_t$ WN.

Note that the d parameter may be estimated by means of the following regression

$$\ln P_x(f_j) = -d\ln 4\sin^2(\pi f_j) + 2\ln\sigma_w, \tag{3.106}$$

for $j = 1, 2, \ldots, z(N)$, in which $z(N) = N^{\alpha}, 0 < \alpha < 1$ (N denotes the number of samples).

Geweke and Porter-Hudak have shown that if $P_x(f_j)$ is estimated by the periodogram method, then the minimum square estimator $\hat{d}$ using the regression (3.106) is normally distributed in big samples if $z(N) = N^{\alpha}$ with $0 < \alpha < 1$:

$$\hat{d} \sim \mathcal{N}\left(d, \frac{\pi^2}{6\sum_{j=1}^{z(N)}(U_j - \bar{U})^2}\right), \tag{3.107}$$

in which $U_j = \ln 4 \sin^2(\pi f_j)$ and $\bar{U}$ is the sample mean of U_j, $j = 1, 2, \ldots, z(N)$.

Under the null hypothesis that there is no LRD ($d = 0$), the t statistics

$$t_{d=0} = \hat{d}\left(\frac{\pi^2}{6\sum_{j=1}^{z(N)}(U_j - \bar{U})^2}\right)^{-1/2} \tag{3.108}$$

has normal distribution in the limit.

3.6 Some H and d estimation methods

3.6.1 R/S statistics

The log-log plot of the R/S statistics *versus* N, in which N denotes the series number of points, of a LRD series is close to straight line with slope $1/2 < H < 1$. First, we evaluate the R/S statistics using N_1 consecutive observations of the series, in which N_1 must be a sufficiently large number. Next, we increase the number of observations by a factor f; i. e., we evaluate R/S over $N_i = f N_{i-1}$ consecutive samples for $i = 2, \ldots, s$. Note that to obtain the R/S statistics with N_i consecutive observations, we can divide the series in N/N_i blocks and obtain N/N_i values, in which . denotes the integer part of a real number.The regression of the log-log plot of all R/S statistics *versus* N_i, $i = 1, \ldots, s$, produces an estimate of the H parameter [9, 128].

3.6.2 Variance plot

The variance plot is a heuristical estimation method of the Hurst parameter. Beran [9] shows that the sample mean variance of a LRD series decreases with its size N slower than in the traditional case (independent or non-correlated variables) as

$$\mathrm{Var}(\bar{x}) \approx cN^{2H-2}, \tag{3.109}$$

in which $c > 0$.

We have the following steps:

1. Let k be an integer number. For different k in the range $2 \leq k \leq N/2$, and for a sufficient number m_k of k-size subseries, evaluate the means of m_k samples of size k, $\bar{x}_1(k), \bar{x}_2(k), \ldots, \bar{x}_{m_k}(k)$ and the global mean

$$\bar{x}(k) = m_k^{-1} \sum_{j=1}^{m_k} \bar{x}_j(k). \tag{3.110}$$

2. For each k, evaluate the sample variance of m_k sample means $\bar{x}_j(k)$, $j = 1, 2, \ldots, m_k$:

$$s^2(k) = \frac{1}{m_k - 1} \sum_{k=1}^{m_k} (\bar{x}_j(k) - \bar{x}(k))^2. \tag{3.111}$$

3. Represent in a plot $\log s^2(k)$ *versus* $\log k$.

For the short range dependence or independence cases, we expect the plot's angular coefficient $2H - 2$ to be $(2 \times 1/2) - 2 = -1$.

3.6.3 Periodogram method

The SDF of a LRD process is approximated by the expression $C_P|f|^{1-2H}$ when $f \to 0$. As the SDF may be approximated by the periodogram, a periodogram's log-log plot *versus* frequency should follow a straight line with $1 - 2H$ slope for frequencies close to zero. The $\hat{P}_x(f)$ SDF's estimator is obtained by the non-parametric periodogram method [96][9], with data tapering, to reduce the power leakage, and smoothing, to reduce the variability of $\hat{P}_x(f)$. The periodogram is evaluated by[10]

$$\hat{P}_x(f) = \frac{1}{N}|X(f)|^2. \tag{3.112}$$

[9] The spectral analysis parametric methods are based on AR, MA and ARMA models. Therefore, they should not be applied to estimate the SDF of a $1/f^{\alpha}$ noise.

[10] The definition was given without including tapering and smoothing to easy the understanding of the estimator's essential nature.

3.6.4 Whittle's method

The Whittle's estimator is also based on the periodogram and involves minimizing the function [116]

$$Q(\theta) = \int_{-0.5}^{0.5} \frac{\hat{P}_x(f)}{P_x(\theta, f)} df \tag{3.113}$$

in which $\hat{P}_x(f)$ denotes the x_t series' periodogram, $P_x(\theta, f)$ is the theoretical SDF of the ARFIMA(p, d, q) $\boldsymbol{x}_t$ model in the frequency f and $\theta = p, d, q$ represents the unknown parameters vector.

3.6.5 Haslett and Raftery's MV approximate estimator

Consider the ARFIMA model (3.71), rewritten in the format

$$\boldsymbol{x}_t = \Delta^{-d}\phi(B)^{-1}\theta(B)\boldsymbol{w}_t \,. \tag{3.114}$$

Let $\hat{x}_t$ be the optimal 1-step prediction of $\boldsymbol{x}_t$ given the past observations $\mathcal{F}_{t-1} = \{x_{t-1}, x_{t-1}, \ldots, x_1\}$, $e_t = x_t - \hat{x}_t$ is the 1-step prediction error and

$$\zeta = \sigma_w^2, \phi_1, \ldots, \phi_p, d, \theta_1, \ldots, \theta_q \tag{3.115}$$

the parameters vector of the model (3.114).

Harvey [50, p.91] shows that the log-likelihood function of (3.114) is given by an expression known as *prediction's error decomposition*

$$\log L(\zeta)|\mathcal{F}_{t-1} = -\frac{N}{2}\log 2\pi - \frac{N}{2}\log\sigma_w^2 - \frac{1}{2}\sum_{t=1}^{N}\log f_t - \frac{1}{2\sigma_w^2}\sum_{t=1}^{N} e_t^2/f_t \,, \tag{3.116}$$

in which $f_t = \text{Var} e_t/\sigma_w^2$.

Haslett and Raftery [51] proposed a fast procedure to determining an approximation of (3.116), that is used by the program S-PLUS®.

3.6.6 Abry and Veitch's wavelet estimator

Consider a stationary signal $x(t)$, $t \in \mathbb{R}$ and its IDWT

$$x(t) \approx \sum_k u(J,k)\phi_{J,k}(t) + \sum_{j=1}^{J}\sum_k w_{j,k}\psi_{j,k}(t).$$

The stationarity of x_t implies the stationarity of the wavelet coefficients sequences $\{w_{j,k}\}$ in all scales j.

Let $\mathbb{P}_j = E w_{j,k}^2 = \text{Var} w_{j,k}$ be the power of the signal $\{w_{j,k}\}$ or *wavelet variance* in a given scale j [97, p.296]. We can show that [1, 115]

$$\mathbb{P}_j = \int_{\mathbb{R}} |\Psi(\nu)|^2 P_x(\nu/2^j) d\nu \,, \tag{3.117}$$

in which $\Psi(\nu)$, $-\infty < \nu < \infty$, denotes the Fourier transform of the wavelet $\psi(t)$ and $P_x(\nu)$ denotes the SDF of $x(t)$.

Consider $j \to \infty$. Then, $P_x(\nu/2^j)$ may be seen as a magnified version (dilated) of the SDF of $x(t)$ in the low frequencies region ($\nu \to 0$). This being so, we verify that (3.117) is an approximation of the power of $x(t)$ in the low frequencies region for $j \to \infty$ (because the integral (3.117) is weighted by the function $|\Psi(\nu)|^2$).

Assume that $x(t)$ is LRD, i.e.,

$$P_x(\nu) \sim C_P |\nu|^{-\alpha}, \quad \nu \to 0, \quad 0 < \alpha < 1 \,, \tag{3.118}$$

in which $\sim$ means that the ratio between the left and right sides of (3.118) converges to 1.

Equations (3.117) and (3.118) imply that the following expression is valid when $j \to \infty$:

$$\mathbb{P}_j \sim C_P \int_{\mathbb{R}} |\Psi(\nu)|^2 |\nu/2^j|^{-\alpha} d\nu = C_P C 2^{j\alpha} \,, \tag{3.119}$$

in which $C = C(\Psi, \alpha) = \int_{\mathbb{R}} |\Psi(\nu)|^2 |\nu|^{-\alpha} d\nu$.

Equation (3.119) suggests that $H = (1+\alpha)/2$ may be estimated by means of

$$\log_2(\mathbb{P}_j) \sim (2H - 1) j + \text{constant}, \quad j \to \infty \,. \tag{3.120}$$

Given the sampled signal x_k, $k = 0, 1, \ldots, N - 1$, associated to the original signal $x(t)$, we may estimate $\log_2(\mathbb{P}_j)$ using the DWT coefficients $w_{j,k}$, $k = 0, 1, \ldots, N_j - 1$, $j = 1, 2, \ldots, J$, of x_k. The $\log_2(\mathbb{P}_j)$ estimator is given by

$$S_j = \log_2 \left(\frac{1}{N_j} \sum_{k=0}^{N_j - 1} w_{j,k}^2 \right) \approx \log_2(\mathbb{P}_j) \,. \tag{3.121}$$

The set of statistics S_j, $j = 1, 2, \ldots, J$, is called the *wavelet spectrum* of the signal x_k.

The relation (3.120) tells that the wavelet spectrum of x_k is *linear with slope* $\alpha = 2H - 1$ in the dilated time scales. Applying a linear regression between scales j_1 and j_2 of the wavelet spectrum produces the following explicit relation for estimating H [1]:

$$\widehat{H}_{j_1,j_2} = \frac{1}{2}\left[\frac{\sum_{j=j_1}^{j_2} \varepsilon_j j S_j - \sum_{j=j_1}^{j_2} \varepsilon_j j \sum_{j=j_1}^{j_2} \varepsilon_j S_j}{\sum_{j=j_1}^{j_2} \varepsilon_j \sum_{j=j_1}^{j_2} \varepsilon_j j^2 - \left(\sum_{j=j_1}^{j_2} \varepsilon_j j\right)^2} + 1\right] \tag{3.122}$$

in which $\varepsilon_j = (N \ln^2 2)/2^{j+1}$.

Eq. (3.122) defines the Abry and Veitch's (AV) wavelet estimator. The wavelet estimator $\widehat{H}_{j_1,j_2}$ presents a good performance when the series are not very distant from a FGN. Empirical studies indicate that it is robust as far as smooth deterministic trends and short range dependence structure's changes of the time series are concerned [1,7,115].

The discussion of this section has shown that the AV estimator is based on the wavelet variance of continuous time signals. Abry, Veitch and Taqqu [120] proposed a method to enable the AV estimator to be applied to discrete time signals, as the networks traffic signals. The method consists on pre-filtering the original discrete time signal x_k (DWT initialization phase), that produces the initial sequence to be decomposed by the DWT

$$u_{\tilde{x}}(k) = \int_{-\infty}^{\infty} \tilde{x}_t \phi_0(t-k)\, dt\,, \tag{3.123}$$

in which

$$\tilde{x}(t) = \sum_{k=-\infty}^{\infty} x(k)\mathrm{sinc}(t-k)\,. \tag{3.124}$$

$\tilde{x}(t)$ is a “fictious” continuous time signal that the DWT initialization procedure associates to the original signal x_k and (3.123) shows that the AV estimator is evaluated by means of the DWT of a filtered version of x_k.

Applying (3.124) in (3.123) we have

$$\begin{aligned} u_{\tilde{x}}(k) &= \int_{-\infty}^{\infty} \tilde{x}_t \phi_0(t-k)\,dt \\ &= \sum_{k=-\infty}^{\infty} x(k) \int_{-\infty}^{\infty} \phi_0(t-k)\mathrm{sinc}(t-k)\,dt \\ &= \sum_{k=-\infty}^{\infty} x(k) I(k-n) \\ &= x(k) \star I(k)\,, \end{aligned} \tag{3.125}$$

in which

$$I(m) = \int_{-\infty}^{\infty} \mathrm{sinc}(t+m)\phi_0(t)\,dt\,. \tag{3.126}$$

Figure 3.19 illustrates, from top to bottom, the wavelet spectra of:

- a WGN,
- of the AR(4) $\boldsymbol{x}_t = 2,7607\boldsymbol{x}_{t-1} - 3,8106\boldsymbol{x}_{t-2} + 2,6535\boldsymbol{x}_{t-3} - 0,9238\boldsymbol{x}_{t-4} + \boldsymbol{w}_t$ model and
- of the `BellcoreAug89` trace (*bin* of 10 miliseconds).

The WGN's wavelet spectrum is flat; the `BellcoreAug89` trace's spectrum is approximately linear between scales $j = 3$ and $j = 10$; the *spikes* in scales 2 and 3 of the AR(4) model's spectrum suggest a short range dependence presence.

Figures 3.20 and 3.21 show the smoothed periodograms by the WOSA method of the AR(4) and Bellcore traces.

3.7 Bi-spectrum and linearity test

Let $\boldsymbol{x} = \boldsymbol{x}_1, \boldsymbol{x}_2, \ldots, \boldsymbol{x}_n{}^{\mathrm{T}}$ and $\omega = \omega_1, \omega_2, \ldots, \omega_n{}^{\mathrm{T}}$, in which T denotes the transpose operation, a n-dimensional real random vector and a real parameters vector with n components, respectively.

The joint moments of order $r = k_1 + k_2 + \ldots + k_n$ of $\boldsymbol{x}$ are given by [90]

$$\mathrm{Mom}\{\boldsymbol{x}_1^{k_1}, \boldsymbol{x}_2^{k_2}, \ldots, \boldsymbol{x}_n^{k_n}\} \equiv E\{\boldsymbol{x}_1^{k_1}\boldsymbol{x}_2^{k_2} \ldots \boldsymbol{x}_n^{k_n}\} = (-j)^r \left. \frac{\partial^r \Phi_{\boldsymbol{x}}(\omega^{\mathrm{T}})}{\partial\omega_1^{k_1}\partial\omega_2^{k_2} \ldots \partial\omega_n^{k_n}} \right|_{\omega_1=\omega_2=\ldots=\omega_n=0} \tag{3.127}$$

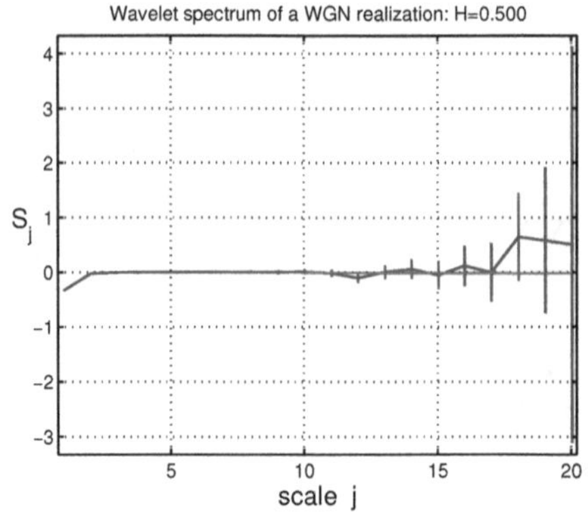

(a)

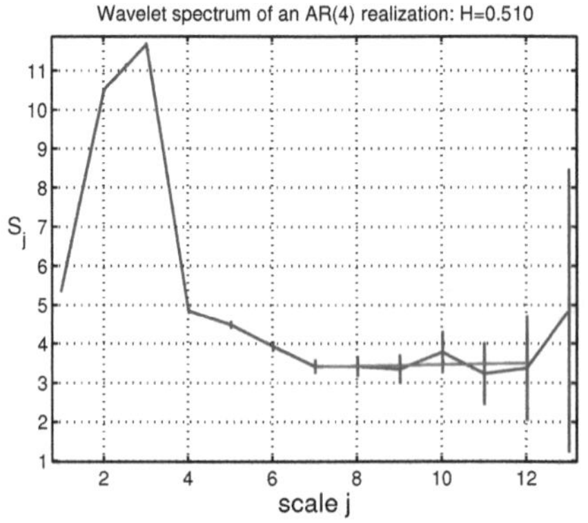

(b)

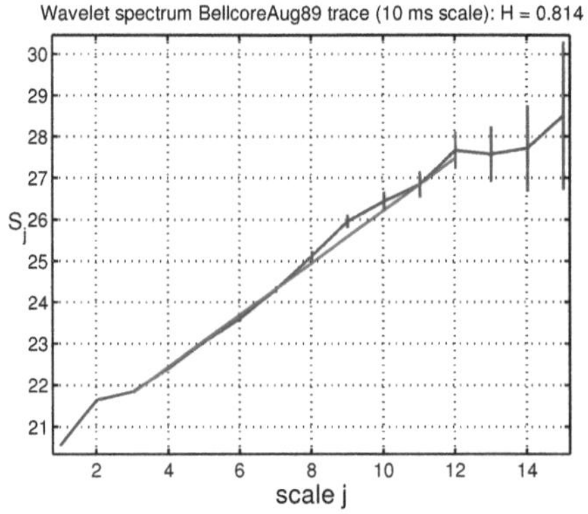

(c)

Figure 3.19: From top to bottom, the wavelet spectra of a WGN, of the AR(4) $\boldsymbol{x}_t = 2,7607\boldsymbol{x}_{t-1} - 3,8106\boldsymbol{x}_{t-2} + 2,6535\boldsymbol{x}_{t-3} - 0,9238\boldsymbol{x}_{t-4} + \boldsymbol{w}_t$ model and of the `BellcoreAug89` trace (*bin* of 10 miliseconds).

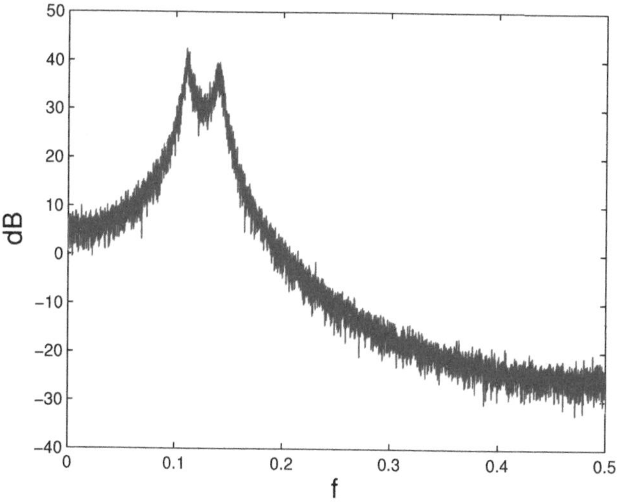

Figure 3.20: Smoothed periodogram by the WOSA method: AR(4)

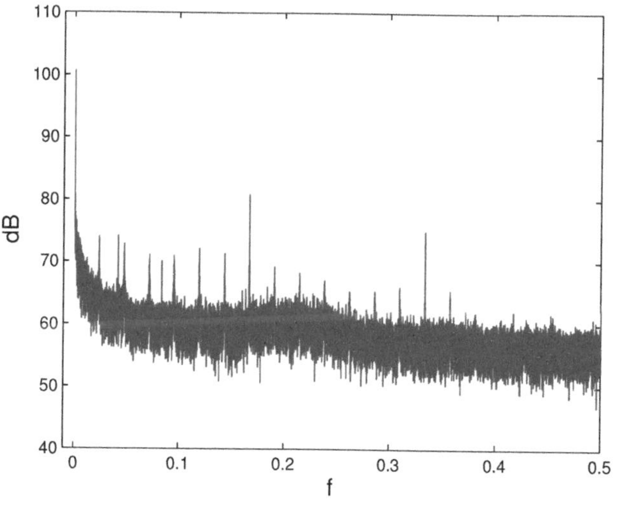

Figure 3.21: Smoothed periodogram by the WOSA method: Bellcore trace

in which

$$\Phi_x(\omega^{\mathrm{T}}) \equiv E\{e^{j\omega^{\mathrm{T}}x}\} \tag{3.128}$$

is the joint characteristic function of x.

The joint cumulants of order r of x are defined as [18, 107]

$$\mathrm{Cum}\{x_1^{k_1}, x_2^{k_2}, \ldots, x_n^{k_n}\} \equiv (-j)^r \left. \frac{\partial^r \tilde{\Psi}_x(\omega^{\mathrm{T}})}{\partial\omega_1^{k_1}\partial\omega_2^{k_2}\ldots\partial\omega_n^{k_n}} \right|_{\omega_1=\omega_2=\ldots=\omega_n=0}, \tag{3.129}$$

in which

$$\tilde{\Psi}_x(\omega^{\mathrm{T}}) \equiv \ln \Phi_x(\omega^{\mathrm{T}}) \tag{3.130}$$

corresponds to the natural logarithm of the joint characteristic function.

We can verify that the moments [18]

$$m_1 = E\{x\} = \mu \qquad m_2 = E\{x^2\} \tag{3.131}$$

$$m_3 = E\{x^3\} \qquad m_4 = E\{x^4\} \tag{3.132}$$

of a random variable x are related to its cumulants by

$$c_1 = \mathrm{Cum}\{x\} = m_1 \tag{3.133}$$

$$c_2 = \mathrm{Cum}\{x, x\} = m_2 - m_1^2 \tag{3.134}$$

$$c_3 = \mathrm{Cum}\{x, x, x\} = m_3 - 3m_2m_1 + 2m_1^3 \tag{3.135}$$

$$c_4 = \mathrm{Cum}\{x, x, x, x\} = m_4 - 4m_3m_1 - 3m_2^2 + 12m_2m_1^2 - 6m_1^4. \tag{3.136}$$

Assume the moments of a stationary process $\{x_k\}$ up to order n exist. Then

$$\begin{aligned}\mathrm{Mom}\{x(k), x(k+\tau_1), \ldots, x(k+\tau_{n-1})\} &= \\ E\{x(k)x(k+\tau_1)\ldots x(k+\tau_{n-1})\} &\equiv \\ m_n^x(\tau_1, \tau_2, \ldots, \tau_{n-1})\end{aligned} \tag{3.137}$$

in which $\tau_1, \tau_2, \ldots, \tau_{m-1}, \tau_i = 0, \pm 1, \pm 2, \ldots$ for all i, denote lags.

Similarly, the nth order cumulants of $\{x_k\}$ can be written as

$$c_n^x(\tau_1, \tau_2, \ldots, \tau_{n-1}) \equiv \mathrm{Cum}\{x(k), x(k+\tau_1), \ldots, x(k+\tau_{n-1})\}. \tag{3.138}$$

Nikias *et al* [87] have shown that the following relations are valid:

$$c_1^x = m_1^x = \mu \text{ (média)} \tag{3.139}$$

$$c_2^x(\tau_1) = m_2^x(\tau_1) - (m_1^x)^2 \text{ (autocovariance function)} \tag{3.140}$$

$$c_3^x(\tau_1, \tau_2) = m_3^x(\tau_1, \tau_2) - m_1^x m_2^x(\tau_1) + m_2^x(\tau_2) + m_2^x(\tau_2 - \tau_1) + 2(m_1^x)^3 \tag{3.141}$$

and

$$\begin{aligned} c_4^x(\tau_1,\tau_2,\tau_3) = {} & m_4^x(\tau_1,\tau_2,\tau_3) - m_2^x(\tau_1)m_2^x(\tau_3-\tau_2) - \\ & - m_2^x(\tau_2)m_2^x(\tau_3-\tau_1) - m_2^x(\tau_3)m_2^x(\tau_2-\tau_1) - \\ & - m_1^x m_3^x(\tau_2-\tau_1,\tau_3-\tau_1) + m_3^x(\tau_2,\tau_3) + \\ & + m_3^x(\tau_2,\tau_4) + m_3^x(\tau_1,\tau_2) + \\ & + (m_1^x)^2 m_2^x(\tau_1) + m_2^x(\tau_2) + m_2^x(\tau_3) + \\ & + m_2^x(\tau_3-\tau_1) + m_2^x(\tau_3-\tau_2) + m_2^x(\tau_2-\tau_1) - \\ & - 6(m_1^x)^4 \,. \end{aligned} \tag{3.142}$$

Let $\{\boldsymbol{x}_k\}$ be a null mean process, i. e., $m_1^x = 0$. If $\tau_1 = \tau_2 = \tau_3 = 0$ in (3.140), (3.141), and (3.142) we have

$$\begin{aligned} E\{\boldsymbol{x}(k)^2\} &= c_2^x(0) = \sigma^2 \text{ (variance)} \\ E\{\boldsymbol{x}(k)^3\}/\sigma^3 &= c_3^x(0,0)/\sigma^3 = S(\boldsymbol{x}_k) \text{ (asymmetry)} \\ E\{\boldsymbol{x}(k)^4\}/\sigma^4 &= c_4^x(0,0,0)/\sigma^4 = K(\boldsymbol{x}_k) \text{ (kurtosis)} \,. \end{aligned} \tag{3.143}$$

The expressions (3.143) give the variance, asymmetry and kurtosis in terms of cumulants.

Let $f = f_1, f_2, \ldots, f_n{}^{\mathrm{T}}$ and $\tau = \tau_1, \tau_2, \ldots, \tau_n{}^{\mathrm{T}}$ be normalized frequencies and lags vectors, respectively.

The nth order polyspectrum of the process $\boldsymbol{x}_k$ is defined as [87]:

$$P_n^x(f^{\mathrm{T}}) = \sum_{\tau_1=-\infty}^{\infty} \cdots \sum_{\tau_{n-1}=-\infty}^{\infty} c_n^x(\tau^{\mathrm{T}}) \exp\{-j2\pi(f^{\mathrm{T}}\tau)\} \,. \tag{3.144}$$

The SDF, bi-spectrum and tri-spectrum correspond to the second, third and fourth orders spectra, respectively:

$$P_2^x(f) = P^x(f) = \sum_{\tau_1=-\infty}^{\infty} c_2^x(\tau) e^{-j2\pi f\tau} \,, \tag{3.145}$$

$$P_3^x(f_1,f_2) = \sum_{\tau_1=-\infty}^{\infty} \sum_{\tau_2=-\infty}^{\infty} c_3^x(\tau_1,\tau_2) e^{-j2\pi(f_1\tau_1+f_2\tau_2)} \,, \tag{3.146}$$

$$P_4^x(f_1,f_2,f_3) = \sum_{\tau_1=-\infty}^{\infty} \sum_{\tau_2=-\infty}^{\infty} \sum_{\tau_3=-\infty}^{\infty} c_4^x(\tau_1,\tau_2,\tau_3) e^{-j2\pi(f_1\tau_1+f_2\tau_2+f_3\tau_3)} \,. \tag{3.147}$$

The bi-spectrum and the tri-spectrum are complex.
Another useful statistic is the bicoherence, defined as

$$B_3^x(f_1, f_2) = \frac{P_3^x(f_1, f_2)}{\sqrt{P^x(f_1 + f_2)P^x(f_1)P^x(f_2)}}. \tag{3.148}$$

Hinich [52] developed statistical tests for Gaussianity and linearity. These tests are based on the following properties:

1) if $\{\boldsymbol{x}_k\}$ is Gaussian, its third order cumulants are null; therefore its bi-spectrum is null and
2) if $\{\boldsymbol{x}_k\}$ is linear and non-Gaussian, then its bicoherence is a non-null constant.

3.8 KPSS stationarity test

Stationarity tests assume the null (or working) hypothesis[11] that the series under investigation are of the type $\boldsymbol{x}_t \sim I(0)$ [128] [12].

The KPSS test, proposed by Kwiatkowski, Phillips, Schmidt and Shin is based on the model [68, 128]

$$y_t = \beta^{\mathrm{T}} d_t + \mu_t + u_t \tag{3.149}$$

$$\mu_t = \mu_{t-1} + \boldsymbol{w}_t, \quad \boldsymbol{w}_t \sim \mathcal{N}(0, \sigma_w^2) \tag{3.150}$$

in which β is a parameters vector, d_t is a deterministic components vector (constant or constant plus trend) and u_t is $I(0)$.

Note that μ_t is a random walk.

The null hypothesis that $\boldsymbol{y}_t$ is $I(0)$ is formulated as $H_0 \quad \sigma_w^2 = 0$, which implies that μ_t is a constant. The KPSS statistics for the test of $\sigma_w^2 = 0$ against the alternative hypothesis $\sigma_w^2 > 0$ is given by

$$KPSS = \left(N^{-2} \sum_{t=1}^{N} \hat{S}_t^2\right) / \hat{\lambda}^2 \tag{3.151}$$

in which N is the size of the sample, $\hat{S}_t = \sum_{j=1}^{t} \hat{u}_j$, $\hat{u}_t$ is the residual of a regression of y_t over d_t and $\hat{\lambda}^2$ is a long run variance estimate of u_t using $\hat{u}_t$.

---

11 It is the hypothesis we wish to reject.

12 Unit roots test as the Dickey-Fuller test work with the null hypothesis that the series is $I(1)$.

Under the null hypothesis that y_t is I_0, it can be shown that $KPSS$ converges to a Brownian movement function that depends on the shape of the deterministic trend terms d_t and, on the other hand, is independent of the β vector.

4

State-space modeling

This chapter introduces a new state space LRD teletraffic model. The new model is a finite-dimensional representation, i. e., truncated, of the ARFIMA process. One of the advantages of the state space modeling is the possibility of using the Kalman filter for on-line prediction of future values of teletraffic signals.

4.1 Introduction

The aggregate data traffic with LRD behavior is frequently modeled by the ARFIMA process [23, 55, 108, 112, 113]. The ARFIMA process achieves modeling the empirical networks traffic auto-correlations for small and large lags, differently from a non-parametric model, as the FGN, that only models the correlations for large lag values [79].

The use of the state space modeling technique along the Kalman filtering [60] has been attracting the interest of the researchers in the financial time series area in the recent years, because the state space approach is flexible and more generic than the classical ARIMA system proposed by Box and Jenkins [33, 83, 118].

Incorporating a state space model in the Kalman filter allows for:

a) an efficient model estimation procedure and
b) filtering, prediction and smoothing of the state vector given the series past observations.

When we consider the problem of on-line traffic prediction, the state space advantage consists in allowing a predictions recursive implementation, by means of the Kalman predictor (being adequate for an on-line prediction scheme).

The state space modeling of an ARFIMA process is not trivial because the LRD state space representation is infinite-dimensional. This being so, the introduction of a finite-dimensional approximation of the ARFIMA model can be justified.

Bhansali and Kokoszka indicate two kinds of time series prediction methods [12]:

- Type-I: in which the predictions are made without the need to estimate the H parameter;
- Type-II: in which it is necessary to first estimate the H parameter of the LRD model.

For the Type-I method, we postulate that a LRD series should be generated by a high order AR model and the predictions are obtained by means of a Kalman filter. For the Type-II method, we assume that a LRD series is modeled by the ARFIMA process and we develop a state space finite-dimensional representation named TARFIMA (*Truncated* ARFIMA). Multiple steps predictions are also obtained by means of the Kalman filter.

The literature registers only a few papers about aggregate teletraffic Kalman prediction [67, 71, 127]. However, none of them considers the use of LRD traffic models, confirming the originality of this approach. In fact, the TARFIMA is a new LRD traffic model that enables on-line heterogenous teletraffic prediction schemes [30]. The TARFIMA's difference is its possibility of practical implementation and, in almost all cases, to be as good as the infinite memory models.

4.2 TARFIMA model

Consider the ARFIMA(p, d, q) model, that may be rewritten as

$$\frac{(1-B)^d \phi(B)}{\theta(B)} \boldsymbol{x}_t = \boldsymbol{w}_t. \tag{4.1}$$

As the polynomial $\frac{(1-B)^d \phi(B)}{\theta(B)}$ is infinite, the model (4.1) may be written in terms of an AR(∞). In practice, the prediction obtained starting from an AR(∞) representation is made by truncating the AR(∞) to an AR(L) model, in which L is a sufficiently large value (this value is obtained empirically, as it will be shown later), i. e., L is a value that results in a TARFIMA good approximation:

$$\frac{(1-B)^d \phi(B)}{\theta(B)} \approx 1 - \psi_{1B} - \psi_{2B}^2 - \ldots - \psi_{LB}^L. \tag{4.2}$$

Therefore, a truncated stationary representation AR(L) of an ARFIMA model $\boldsymbol{x}_t$ satisfies the differences equation

$$\psi(B)\boldsymbol{x}_t \approx \boldsymbol{w}_t, \tag{4.3}$$

in which $\psi(B) = 1 - \psi_{1B} - \psi_{2B}^2 - \ldots - \psi_{LB}^L$.

All time series linear model has a state space representation [83, p.367]. The general Gaussian linear state space model may be written as the system of equations [118, p. 508]

$$\underset{m\times 1}{\boldsymbol{x}_{t+1}} = \underset{m\times 1}{d_t} + \underset{m\times m}{T_t} \, . \, \underset{m\times 1}{\boldsymbol{x}_t} + \underset{m\times r}{H_t} \, . \, \underset{r\times 1}{\eta_t}, \tag{4.4}$$

$$\underset{l\times 1}{\boldsymbol{y}_t} = \underset{l\times 1}{c_t} + \underset{l\times m}{Z_t} \, . \, \underset{m\times 1}{\boldsymbol{x}_t} + \underset{l\times 1}{\varepsilon_t}, \tag{4.5}$$

in which $t = 1, 2, \ldots, N$ and

$$\boldsymbol{x}_1 \sim \mathcal{N}(\mu_{1|0}, \Sigma_{1|0}) \quad \text{(initial state)}, \tag{4.6}$$

$$\eta_t \sim \mathcal{N}(0, I_r) \quad \text{(WGN)}, \tag{4.7}$$

$$\varepsilon_t \sim \mathcal{N}(0, I_l) \quad \text{(WGN)}. \tag{4.8}$$

and we assume that

$$E\varepsilon_t\eta_t^{\mathrm{T}} = \mathbf{0}. \tag{4.9}$$

The *state or transition equation* (4.4) describes the time evolution of the state vector $\boldsymbol{x}_t$ using a first order Markovian structure.

The *observation equation* describes the observation vector $\boldsymbol{y}_t$ in terms of the state vector $\boldsymbol{x}_t$ and of a vector ε_t (WGN noise measurement) with covariance matrix I_l.

We assume that the WGN η_t (innovation) with covariance matrix I_r in (4.4) is not correlated with ε_t.

The deterministic matrices T_t, H_t, Z_t are called *system matrices* and the vectors d_t e c_t contain fixed components and may be used to incorporate known effects or patterns in the model (otherwise they are zero). For univariate models, $l = 1$ and, consequently, Z_t is a row vector.

There are several ways of mapping (or transforming) the truncated AR model

$$\boldsymbol{y}_t = \phi_1\boldsymbol{y}_{t-1} + \ldots + \phi_L\boldsymbol{y}_{t-L} + \xi_t, \tag{4.10}$$

in which ξ_t denotes the innovation in a state space format [118].

In this book, we adopt the transformation proposed by Harvey [50]. In this way, (4.10) may be put in a state space format, that will be called *TARFIMA state space model*, with the state and observation equations

$$\boldsymbol{x}_{t+1} = T\boldsymbol{x}_t + H\xi_t, \quad \xi_t \sim \mathcal{N}(0, \sigma_\xi^2) \tag{4.11}$$

$$\boldsymbol{y}_t = Z\boldsymbol{x}_t, \tag{4.12}$$

in which

$$T = \begin{pmatrix} \phi_1 & 1 & 0 & \dots & 0 \\ \phi_2 & 0 & 1 & & 0 \\ \vdots & & & \ddots & \vdots \\ \phi_{L-1} & 0 & 0 & & 1 \\ \phi_L & 0 & 0 & \dots & \end{pmatrix}, \quad H = \begin{pmatrix} 1 \\ 0 \\ \vdots \\ 0 \end{pmatrix}, \tag{4.13}$$

$$\text{and } Z = (1 \quad 0 \quad \dots \quad 0 \quad 0). \tag{4.14}$$

The state vector has the format

$$\boldsymbol{x}_t = \begin{pmatrix} \boldsymbol{y}_t \\ \phi_2 \boldsymbol{y}_{t-1} + \dots + \phi_L \boldsymbol{y}_{t-L+1} \\ \phi_3 \boldsymbol{y}_{t-1} + \dots + \phi_L \boldsymbol{y}_{t-L+2} \\ \vdots \\ \phi_L \boldsymbol{y}_{t-1} \end{pmatrix}. \tag{4.15}$$

4.2.1 Multistep prediction with the Kalman filter

The linear estimation theory states that the MMSE estimate $\hat{\boldsymbol{x}}_t$ of the state $\boldsymbol{x}_t$ in (4.11), given the past observations

$$\mathcal{F}_t = y_t, y_{t-1}, \dots, y_1,$$

corresponds to the conditional mean $E\{\boldsymbol{x}_t|\mathcal{F}_t\}$, denoted by $x_{t|t}$ [114].

The objective of the Kalman filter for the state space defined by (4.11) and (4.12) is to obtain recursive estimates of the conditional mean $x_{t+1|t}$ and of the covariance matrix $\Sigma_{t+1|t}$ of the state vector $\boldsymbol{x}_{t+1}$ given $\mathcal{F}_t$ and the model, i. e.,

$$x_{t+1|t} = E\{T\boldsymbol{x}_t + H\xi_t|\mathcal{F}_t\} = Tx_{t|t}, \tag{4.16}$$

$$\Sigma_{t+1|t} = \text{Var}\{T\boldsymbol{x}_t + H\xi_t|\mathcal{F}_t\} = T\Sigma_{t|t}T^{\text{T}} + \sigma_\xi^2 HH^{\text{T}}. \tag{4.17}$$

Consider

- $y_{t|t-1}$, the conditional mean of $\boldsymbol{y}_t$ given $\mathcal{F}_{t-1}$,
- $\tilde{\boldsymbol{y}}_t = (\boldsymbol{y}_t - y_{t|t-1})$, the 1-step prediction error of $\boldsymbol{y}_t$ given $\mathcal{F}_{t-1}$,
- $\sigma_t^2 = \text{Var}\{\tilde{\boldsymbol{y}}_t | \mathcal{F}_{t-1}\}$, the 1-step prediction error variance,
- $C_t = \text{Cov}(\boldsymbol{x}_t, \tilde{\boldsymbol{y}}_t | \mathcal{F}_{t-1})$, and
- $K_t = TC_t/\sigma_t^2 = T\Sigma_{t|t-1}Z^{\mathrm{T}}/\sigma_t^2$, the Kalman's gain.

The equations of the Kalman filter algorithm are given by [33, 50, 118]

$$\begin{cases} \tilde{\boldsymbol{y}}_t & = \boldsymbol{y}_t - Z\boldsymbol{x}_{t|t-1}, \\ \sigma_t^2 & = Z\Sigma_{t|t-1}Z^{\mathrm{T}}, \\ K_t & = T\Sigma_{t|t-1}Z^{\mathrm{T}}/\sigma_t^2, \\ L_t & = T - K_t Z, \\ \boldsymbol{x}_{t+1|t} & = T\boldsymbol{x}_{t|t-1} + K_t\sigma_t^2, \\ \Sigma_{t+1|t} & = T\Sigma_{t|t-1}L_t^{\mathrm{T}} + \sigma_\xi^2 HH^{\mathrm{T}}, \quad t = 1, \ldots, N. \end{cases} \tag{4.18}$$

As the state space model given by (4.11) and (4.12) is stationary, it can be proved that the $\Sigma_{t|t-1}$ matrices converge to a constant matrix Σ_*. The quantities σ_t^2, K_t, and $\Sigma_{t+1|t}$ become constant when the steady state regime is reached.

Let t be the origin of the prediction and that we want to predict $\boldsymbol{y}_{t+h}$ for $h = 1, 2, \ldots, H$. The prediction equations of the Kalman filter (the last two equations of (4.18)) produce 1-step predictions. In general, the predictions $\hat{y}_h$, $h = 1, \ldots, H$, may be obtained by means of the prediction equation of the Kalman filter extending the data set $\mathcal{F}_t$ with a set of missing values (i. e., the 1-step prediction is used to get the 2-steps prediction and so on).

4.2.2 The prediction power of the TARFIMA model

As it has already been seen, autocorrelations that decay slowly towards zero make it difficult the estimation of constants as the mean of a stationary process. On the other hand, the foreseeability degree of a series increases as the time dependence between observations increases [9]. Therefore, good short and long term predictions may be obtained for LRD processes when we can access a large number of past samples and if the estimators of the p, d e q model parameters are consistent and have low bias. The basic idea of the TARFIMA model is to make a good usage of the LRD series autocorrelation structure.

We adopt the metrics

$$P_h^2 = \frac{\sigma_x^2 - EQM_h}{\sigma_x^2} = 1 - \frac{EQM_h}{\sigma_x^2}. \tag{4.19}$$

EQM_h denotes the MSE of the h-step prediction (or the error variance of the h-step prediction, as $EQM_h = E(\mathbf{e}_{t+h})^2 = E(\boldsymbol{x}_{t+h} - \hat{\boldsymbol{x}}_{t+h})^2)$, that is a measure of the foreseeability of a given model. For a perfect prediction, $EQM_h = 0$ and $P_h^2 = 1$.

On the other hand, if $\boldsymbol{x}_{t+h}$ does not depend on the past samples, then the best prediction is the series non-conditional mean. In this case, the prediction does not improve when past observations are used. Therefore, $EQM_h = \sigma_x^2$ and $P_h^2 = 0$.

Note that $P_h^2 \to 0$ when $h \to \infty$. This reflects the fact that very long term predictions are increasingly difficult.

Consider the ARFIMA$(0, d, 0)$ model

$$\boldsymbol{x}_t = (1 - B)^{-d}\boldsymbol{w}_t$$

in which the innovation $\boldsymbol{w}_t$ has unit power ($\sigma_w^2 = 1$).

In this case, we can demonstrate that the variance of $\boldsymbol{x}_t$ is given by [53]:

$$\sigma_x^2 = \frac{\Gamma(1 - 2d)}{\Gamma^2(1 - d)}. \tag{4.20}$$

We can also demonstrate that the variance of $\boldsymbol{x}_t$ can be evaluated by the alternative expression [20, p.91]

$$\sigma_x^2 = 1 + \sum_{j=1}^{\infty} \psi_j^2, \tag{4.21}$$

in which ψ_j^2 denote the coefficients of the MA(∞) representation of $\boldsymbol{x}_t$.

Note that the expression (4.21) may be applied to an approximate evaluation of the TARFIMA(L) model's variance. To do so, we have to get a truncated representation MA(L') of TARFIMA(L), in which L' is a sufficiently high order (100, for example).

Numerical values of P_h^2, $h = 1, 10, 20, 100$, are provided in Tables 4.1 and 4.2.

Table 4.1 shows the values of P_h^2 for AR(1) and TARFIMA(L), $L = \{1, 10\}$ models. The lag-1 correlation values of the AR(1) model

$$\boldsymbol{x}_t - \phi_1 \boldsymbol{x}_{t-1} = \boldsymbol{w}_t$$

were chosen in such a way that $\rho_1 = 1/9$ and $\rho_1 = 2/3$ correspond to the lag-1 correlations of the ARFIMA(0; 0, 1; 0) and ARFIMA(0; 0, 4; 0) processes, respectively. The optimal prediction of $\boldsymbol{x}_{t+h}$ given $\mathcal{F}_t = \{x_t, x_{t-1}\}$, i. e., the prediction obtained by using the AR(1) model is given by

$$\hat{x}_{t+h} = (\rho_x(1))^h x_t = (\phi_1)^h x_t \tag{4.22}$$

and the h-step prediction error variance by [9, p.168]

$$EQM_h^{AR(1)} = \sigma_x^2 1 - (\rho_x(1))^{2h}. \tag{4.23}$$

Table 4.1: P_h^2 values for AR(1) and TARFIMA(L), $L = \{1, 10\}$ models.

h	AR(1)		$L=1$		$L=10$	
	$\rho_1 = 1/9$	$\rho_1 = 2/3$	$d = 0.1$	$d = 0.4$	$d = 0.1$	$d = 0.4$
1	0.0123	0.4444	0.01	0.16	0.0172	0.3249
10	0	0.0003	0	0	0.0002	0.0380
20	0	0	0	0	0	0.0038
100	0	0	0	0	0	0

Table 4.2: P_h^2 for TARFIMA(L), $L = \{100, 150\}$ and ARFIMA$(0, d, 0)$ (column in which $L = \infty$) models.

h	$L=100$		$L=150$		$L=\infty$	
	$d = 0.1$	$d = 0.4$	$d = 0.1$	$d = 0.4$	$d = 0.1$	$d = 0.4$
1	0.0188	0.4095	0.0189	0.4161	0.0191	0.5169
10	0.0019	0.1585	0.0020	0.1680	0.0022	0.3116
20	0.0009	0.1082	0.0010	0.1182	0.0013	0.2704
100	0	0.0167	0.0001	0.0277	0.0003	0.1955

4.3 Series exploratory analysis

In this section, we make the exploratory analysis of two approximate realizations of the process ARFIMA$(0, d, 0)$, $d = \{0.3; 0.4\}$, two MWM(H), $H = \{0.8; 0.9\}$ realizations, produced by the wavelet-based LRD signals generator developed by Mello, Lima, Lipas and Amazonas [28,31] and of the Nile river yearly minimum levels between years 1007 a 1206.

The series d parameter has been estimated by the following methods:

- R/S analysis;

- frequency domain periodogram based least squares regression;
- Whittle;
- MV Haslett-Raftery approximated estimator;
- AV wavelet estimator.

The following statistics or tests have also been used:

- SACF;
- smoothed periodograms by the WOSA $\hat{P}_x^{\mathrm{WOSA}}(f)$ and Daniell's autocorrelation window $\hat{P}_x^{lw}(f)$ methods;
- R/S and GPH long memory statistical tests;
- Bi-spectrum concept based normality and Hinich's linearity tests.

The normality of the series marginal distributions and of the estimated models' residuals have also been investigated by means of histograms and QQ-plot graphs. The QQ-plot is a graph of the standard normal's quantiles on the horizontal axis *vs* the series ordered values on the vertical axis.

$\mathrm{AR}(p)$ (for the Type-I prediction methods implementation) and $\mathrm{ARFIMA}(p, d, q)$ (for the Type-II prediction by means of the $\mathrm{TARFIMA}(L)$ model) models have been estimated. The diagnostics of the adjusted models have been made by SACF and residual's QQ-plot analysis.

4.3.1 ARFIMA(0; 0.4; 0) series

Figure 4.1 shows the simulated ARFIMA(0; 0.4; 0) series.

An ARFIMA(0; 0.3684; 0) model has been estimated by the function `S+FinMetrics` `FARIMA` [128, p.271].

The function `FARIMA` implements a fast estimation procedure of the d parameter based on the approximate MV estimation method proposed by Haslett and Raftery and adjusts the p and q parameters selecting $p \leq p_{\max}$ and $q \leq q_{\max}$ values that minimize the BIC (we used $p_{\max} = q_{\max} = 2$).

Table 4.3 summarizes the obtained results.

Table 4.3: d parameter estimates for the ARFIMA(0; 0.4; 0) realization.

R/S analysis	periodogram with Daniell's window	Whittle estimator	Haslett-Raftery MV estimator	Abry/Veitch wavelet estimator
0.2735	0.3696	0.3773	0.3684	0.3800

Note that the result of the R/S analysis disagrees from the others, that are in the range $0.36 \leq \hat{d} \leq 0.38$.

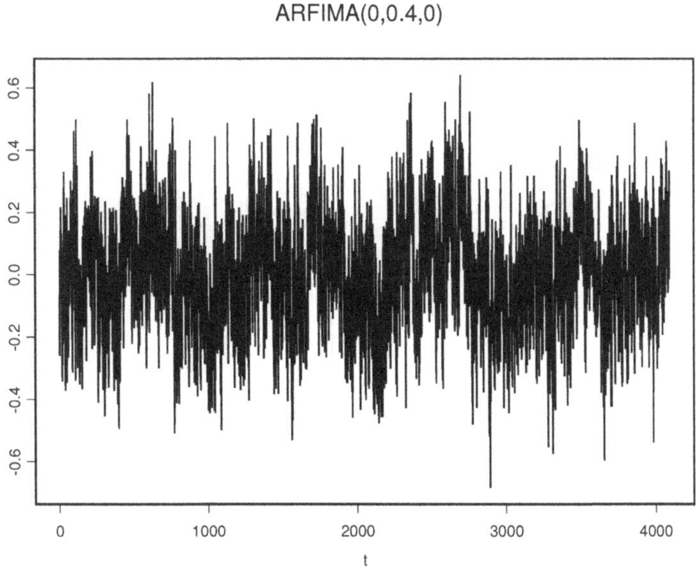

Figure 4.1: Simulation: ARFIMA(0; 0.4; 0).

Table 4.4 shows that the KPSS test cannot reject the null hypothesis (H_0: the series is stationary) because $p_{af} > 0.05$ ("false alarm" probability greater than the significance level $\alpha = 0.05$), as indicated by "YES" in the column KPSS.

The R/S does not reject the null hypothesis of SRD series (H_0: the series does not have long memory), but the GPH test does.

The Hinich tests cannot reject the null hypotheses of normality (H_0: series with null bi-spectrum) and linearity (H_0: series with constant bicoherence).

The number of estimated unit roots (column $\hat{m}$) by the function `S-PLUS`® `FARIMA` is equal to zero.

Figures 4.2, 4.3 and 4.4 show the ARFIMA(0; 0.4; 0) periodogram. The first one is the raw periodogram and the other two are smoothed periodograms obtained by the Daniell's window [96, p.264] and the WOSA method [96, p.289].

In Figures 4.2, 4.3 and 4.4 it is clearly seen the singularity of the $\frac{1}{f^{\alpha}}$-type given that the SDF increases as $f \to 0$.

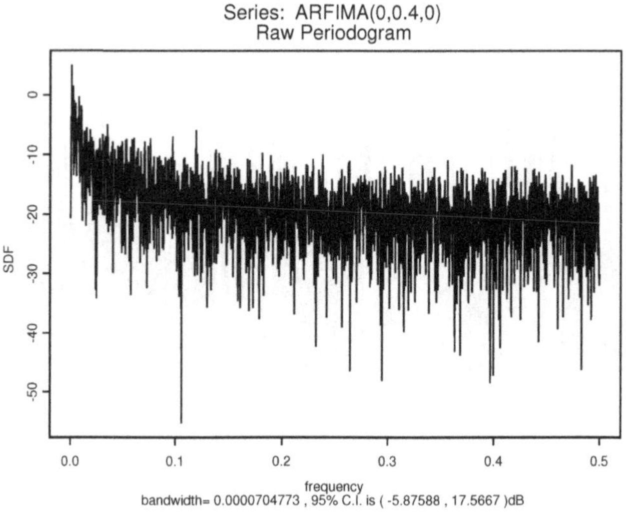

Figure 4.2: ARFIMA(0; 0.4; 0) simulation: periodogram.

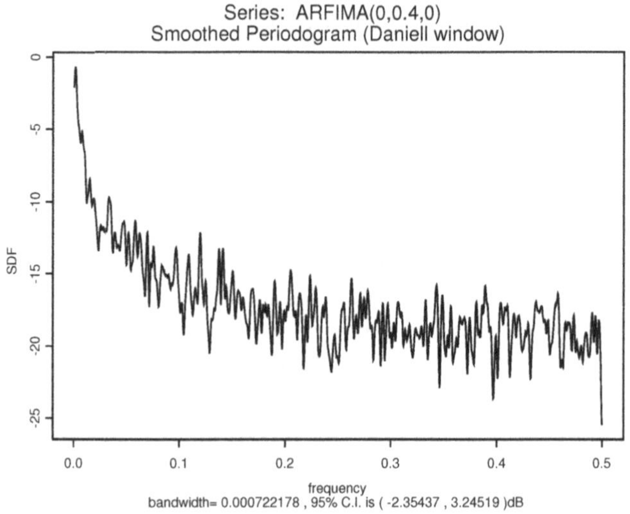

Figure 4.3: ARFIMA(0; 0.4; 0) simulation: Daniell's window smoothed periodogram.

Table 4.4: ARFIMA(0; 0.4; 0) series: stationarity, long memory, normality and linearity tests, and number of unit roots (column $\hat{m}$).

x_t is $I(0)$?	x_t is LRD?		Hinich		$\hat{m}$
KPSS	R/S	GPH	x_t is normal?	x_t is linear?	
YES	NO	YES	YES	YES	0
$p_{af} > 0.05$	$p_{af} > 0.05$	$p_{af} \leq 0.01$	$p_{af} = 0.5213$		

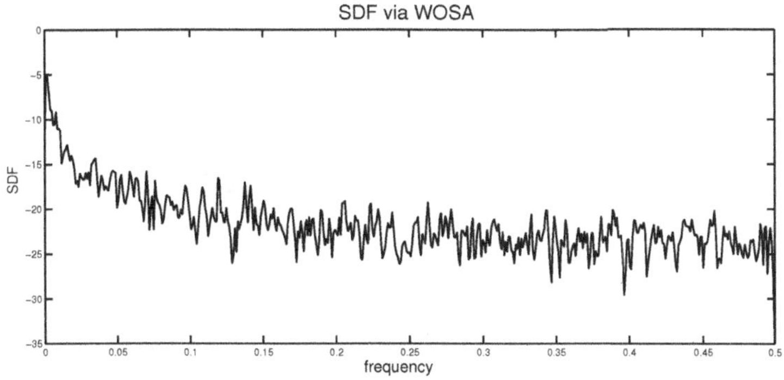

Figure 4.4: ARFIMA(0; 0.4; 0) simulation: estimated periodogram by the WOSA method.

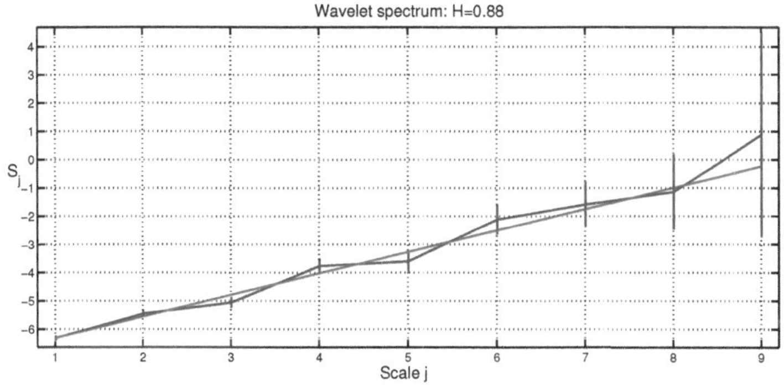

Figure 4.5: Wavelet spectrum of the ARFIMA(0; 0.4; 0) signal.

Figure 4.5 shows the wavelet spectrum of the ARFIMA(0; 0.4; 0) signal. Such linear behavior is a clear indication of LRD.

Figure 4.6 shows the QQ-plot of the ARFIMA(0; 0.4; 0) series. The QQ-plot indicates that the series is Gaussian.

Figure 4.7 shows a graph with the theoretical SACF and ACFs of the estimated ARFIMA(0; 0.3684; 0) and AR(15) models. The AR(15) process is the best AR model according to the AIC criterium and has been estimated by the function `S-PLUS`® `AR` (Yule-Walker method). The figure shows that the AR(15) process does not model the slow decay of the series' SACF.

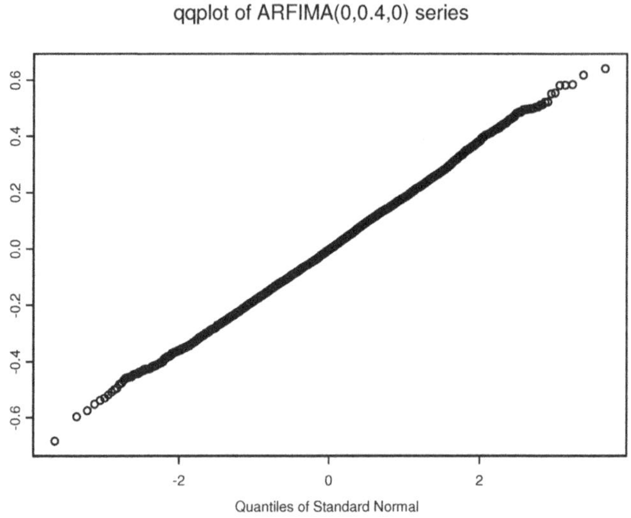

Figure 4.6: QQ-plot.

Figure 4.8 shows the QQ-plot of the residuals and Figure 4.9 shows the residuals' SACF graph. Both figures indicate that the residuals may be taken as WGN.

Figure 4.10 shows the AR(15) model's poles and zeros diagram. The high order adjusted AR model achieves modeling the ARFIMA(0;0.4;0) long memory at the cost of positioning a pole close to the unit circle.

4.3.2 MWM series with $H = 0.9$

Figure 4.11 shows one realization of a MWM($H = 0.9$) series.

An ARFIMA(0; 0.3572; 0) model has been estimated by the function `S+FinMetrics` `FARIMA`. Table 4.5 summarizes the obtained results.

Table 4.6 summarizes the statistical tests results:

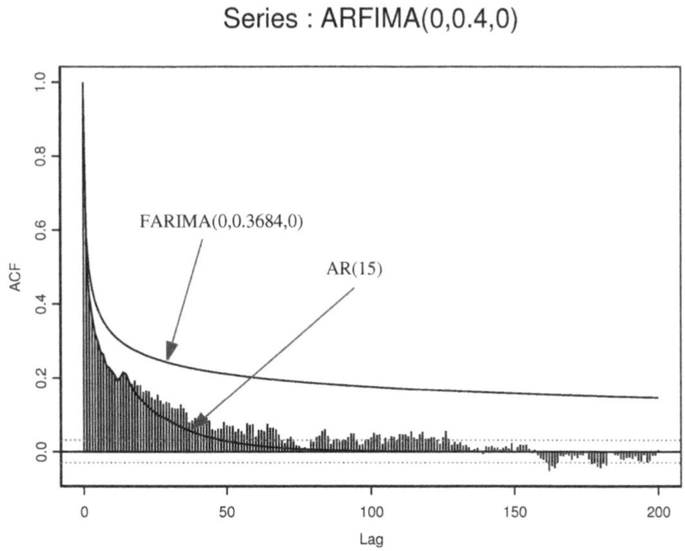

Figure 4.7: Theoretical SACF and ACF graphs of the estimated ARFIMA(0; 0.3684; 0) and AR(15) models.

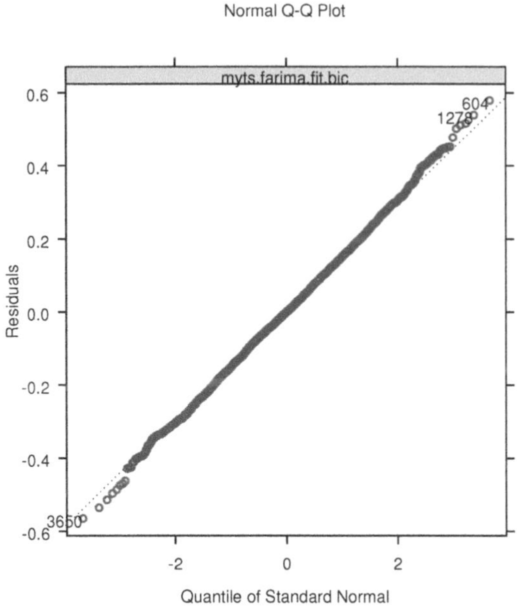

Figure 4.8: Residuals' QQ-plot.

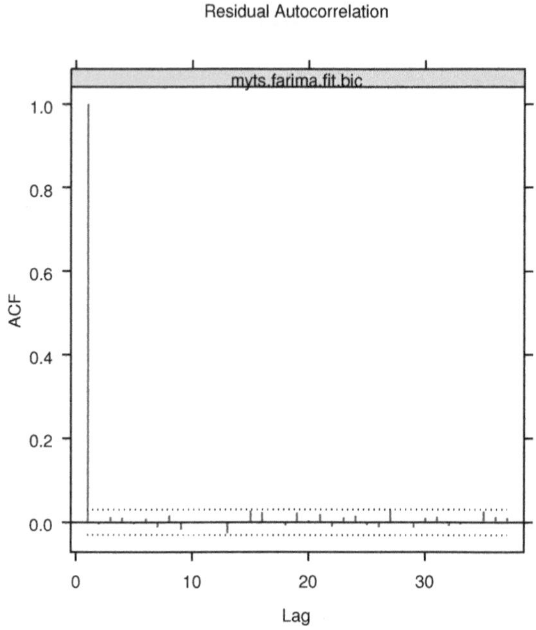

Figure 4.9: Residuals' SACF graph.

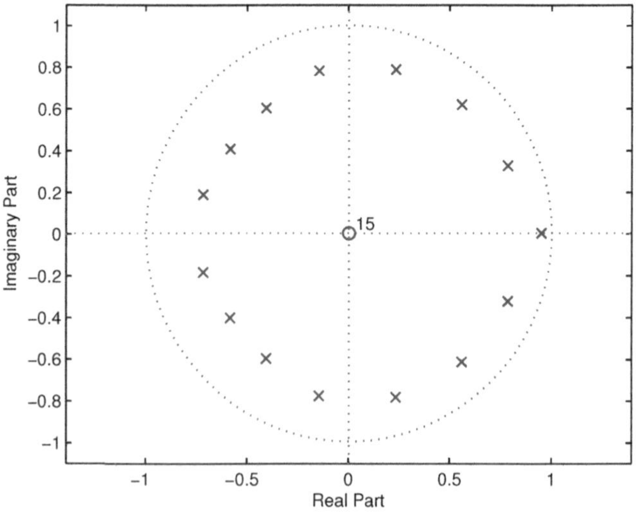

Figure 4.10: AR(15) model's poles and zeros diagram.

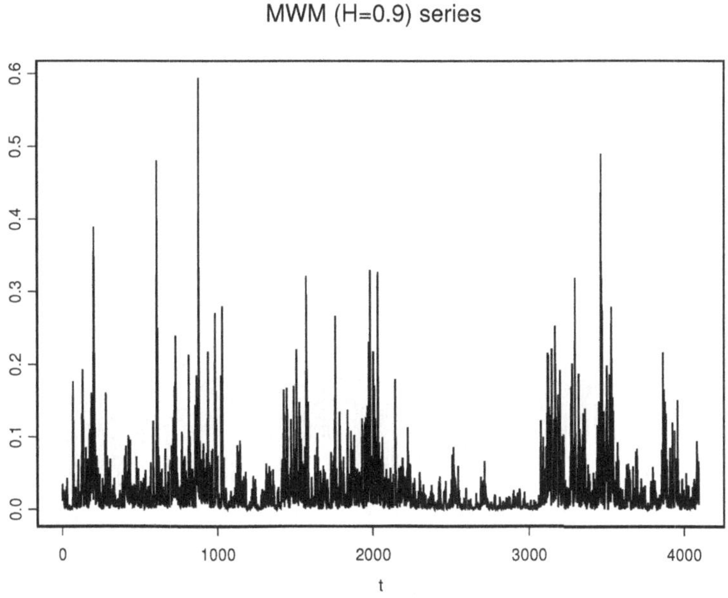

Figure 4.11: Realization: MWM($H = 0.9$).

Table 4.5: H estimates for the MWM($H = 0.9$) realization.

R/S analysis	periodogram with Daniell's window	Whittle estimator	Haslett-Raftery MV estimator	Abry/Veitch wavelet estimator
0.3492	0.3267	0.3428	0.3572	0.362

- the KPSS rejected the stationarity hypothesis;
- both R/S and GPH as well indicate that the process is LRD;
- Hinich tests rejected the hypotheses of normality and linearity;
- the number of unit roots is zero.

Table 4.6: MWM($H = 0.9$) series: statistical tests results and number of unit roots.

x_t é $I(0)$?	x_t é LRD?		Hinich		$\hat{m}$
KPSS	R/S	GPH	x_t is normal?	x_t is linear?	
NO	YES	YES	NO	NO	0
$p_{af} \leq 0.05$	$p_{af} \leq 0.01$	$p_{af} \leq 0.01$	$p_{af} = 0$		

Figures 4.12, 4.13 and 4.14 show the ARFIMA(0; 0.3572; 0) periodogram. The first one is the raw periodogram and the other two are smoothed periodograms obtained by the Daniell's window and the WOSA method.

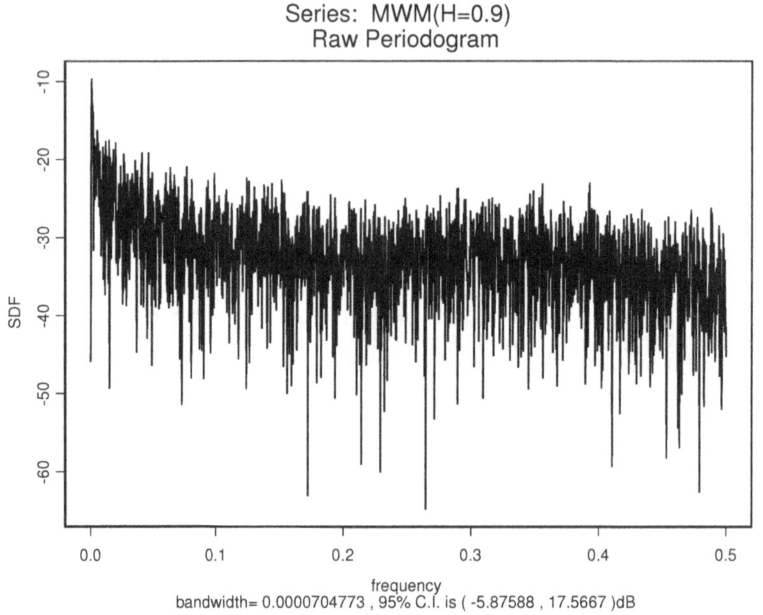

Figure 4.12: MWM($H = 0.9$) simulation: periodogram.

Figure 4.15 shows the QQ-plot of the ARFIMA(0; 0.3572; 0) series. The QQ-plot indicates that the series is not Gaussian.

Figure 4.16 shows a graph with the theoretical SACF and ACFs of the estimated ARFIMA(0; 0.3572; 0) and AR(28) models. The AR(28) process is the best AR model according to the AIC criterium and has been estimated by the function `S-PLUS`® `AR` (Yule-Walker method). The figure shows that the AR(28) process does not model the slow decay of the series' SACF.

Figure 4.17 shows the QQ-plot of the residuals and Figure 4.18 shows the residuals' SACF graph. The QQ-plot shows some deviation from a WGN while the SACF indicates that the residuals may be taken as WGN.

Figure 4.19 shows the AR(28) model's poles and zeros diagram. The high order adjusted AR model achieves modeling the ARFIMA(0;0.3572;0) long memory at the cost of positioning a pole close to the unit circle.

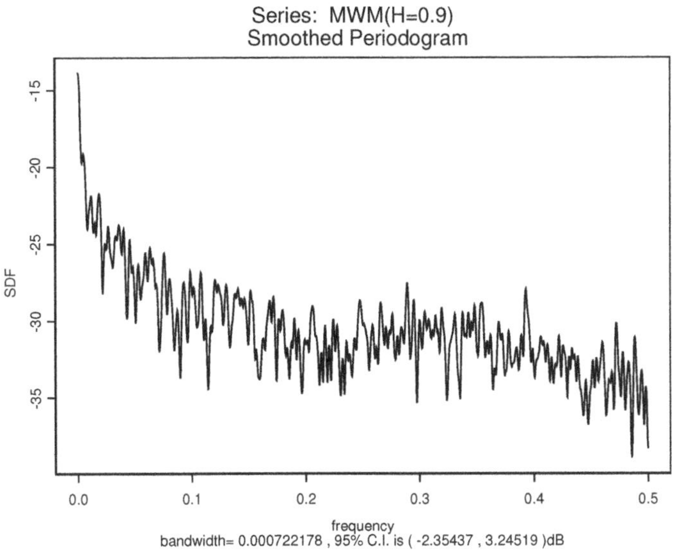

Figure 4.13: MWM($H = 0.9$) simulation: Daniell's window smoothed periodogram.

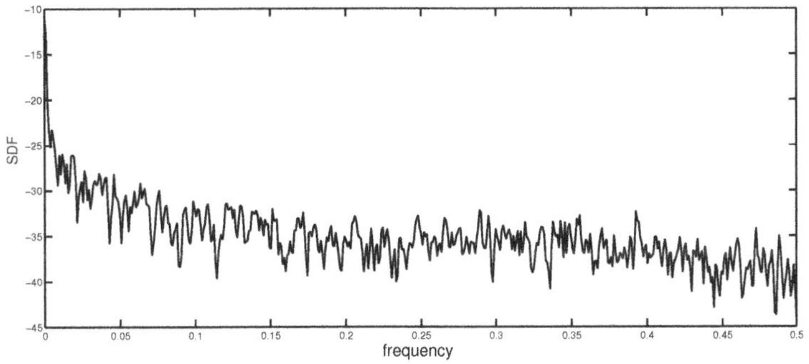

Figure 4.14: MWM($H = 0.9$) simulation: estimated periodogram by the WOSA method.

4.3.3 Nile river series

Figure 4.20 shows the Nile river series between years 1007 and 1206.

According to Beran [9], and Percival and Walden [97, pág.386], the Nile river series between years 622 and 1284 is not globally stationary. Percival and Walden [97] show, using the DWT, that the series behavior changes around year 715 and that it can be subdivided in two locally stationary

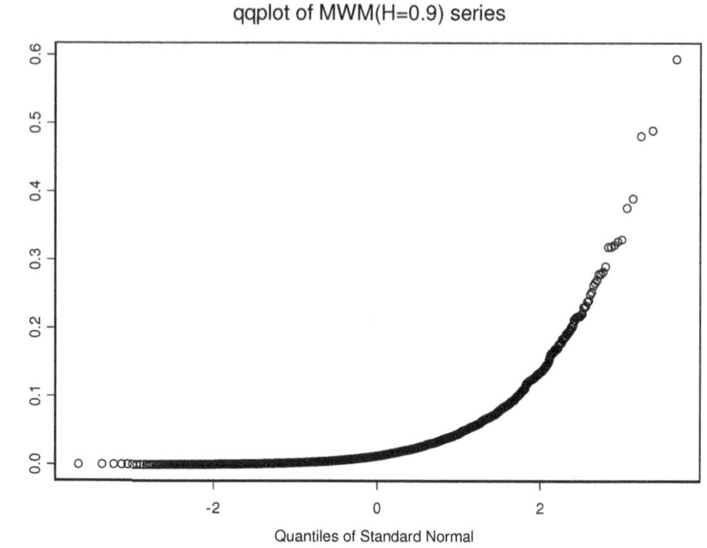

Figure 4.15: MWM(H = 0.9) simulation: QQ-plot.

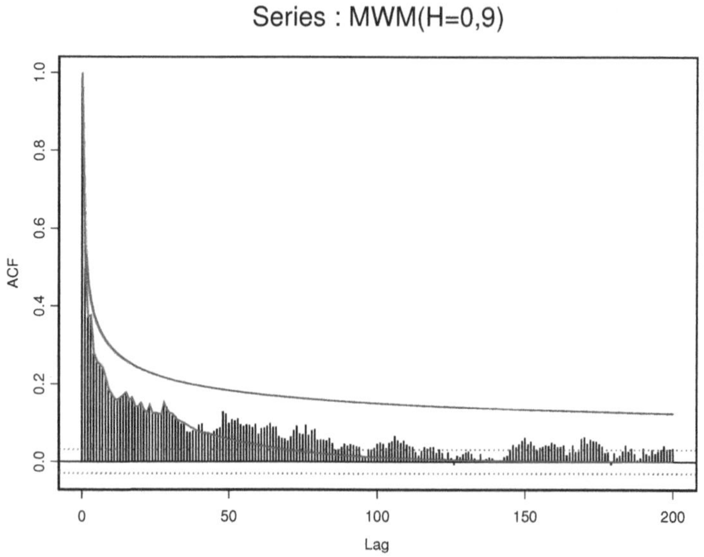

Figure 4.16: MWM(H = 0.9) simulation: Theoretical SACF and ACFs graphs of the estimated ARFIMA(0; 0.3572; 0) (red line) and AR(28) (green line) models.

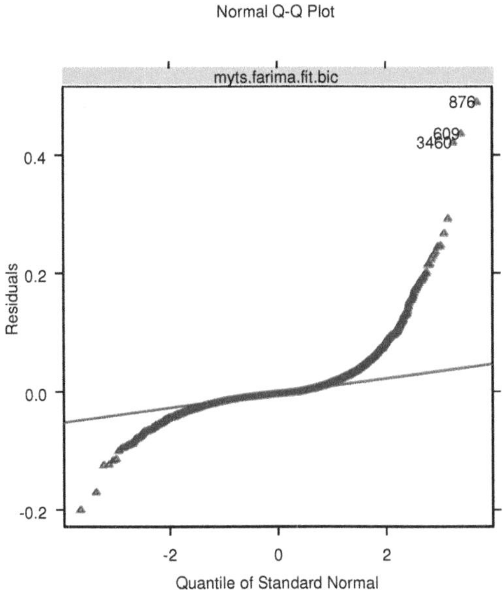

Figure 4.17: Residuals' QQ-plot.

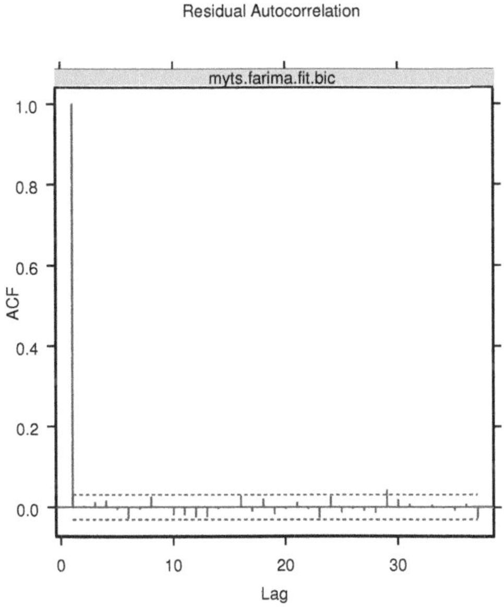

Figure 4.18: Residuals' SACF graph.

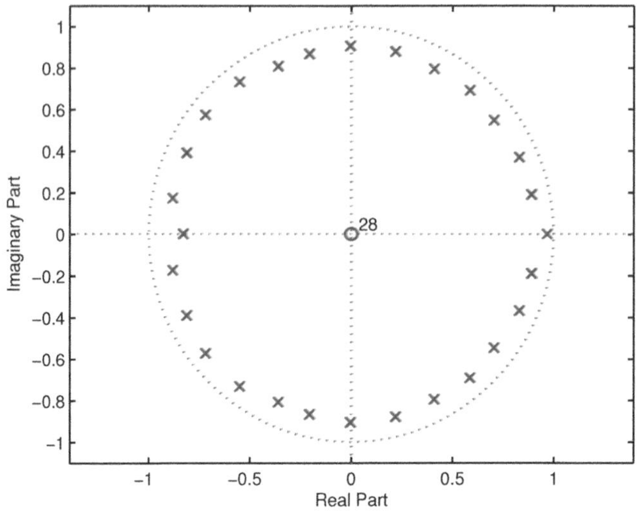

Figure 4.19: AR(28) model's poles and zeros diagram.

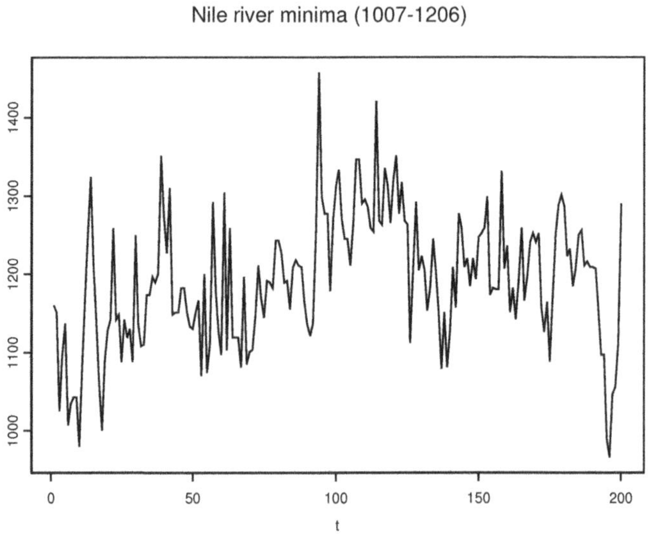

Figure 4.20: Nile river series between years 1007 and 1206.

ARFIMA series. Percival and Walden [97] estimate $\hat{H} = 0.9532$ by the DWT based MV approximate method. The function `S+FinMetrics FARIMA` adjusted an ARFIMA(0; 0.505; 0) model (non-stationary as $d > 0.5$). As the adjusted model by the function `FARIMA` is non-stationary, this section will adopt the ARFIMA(0; 0.4125; 0) model, in which the d parameter value corresponds to estimated value by the Whittle method. Table 4.7 summarizes the Nile river series d parameter estimates.

Table 4.7: Nile river series d parameter estimates.

R/S analysis	periodogram with Daniell's window	Whittle estimator	Haslett-Raftery MV estimator	Abry/Veitch wavelet estimator
0.2853	0.4062	0.4125	0.5050	0.387

Table 4.8 shows the results of the statistical test and the estimated number of unit roots of the Nile river series. The KPSS rejected the series stationarity hypothesis. The series is LRD according to the R/S and GPH tests. The series is linear, but non-Gaussian, according to the Hinich tests. The results of the Hinich tests (linear non-Gaussian series) suggest that the marginal probability distribution of the Nile river series can be modeled by a stable distribution. There are not unit roots.

Table 4.8: Nile river series: stationarity, long memory, normality and linearity tests, and number of unit roots (column $\hat{m}$).

x_t is $I(0)$?	x_t is LRD?		Hinich		$\hat{m}$
KPSS	R/S	GPH	x_t is normal?	x_t is linear?	
NO	YES	YES	NO	YES	0
$p_{af} < 0.01$	$p_{af} < 0.01$	$p_{af} < 0.05$	$p_{af} = 0$		

Figures 4.21, 4.22 and 4.23 show the Nile river series periodogram. The first one is the raw periodogram and the other two are smoothed periodograms obtained by the Daniell's window and the WOSA method. All figures indicate the presence of $\frac{1}{f^\alpha}$ behavior as $f \to 0$.

Figure 4.24 shows the Nile river series wavelet spectrum. The straight line is a clear indication of LRD.

Figure 4.25 shows the QQ-plot of the Nile river series. The QQ-plot indicates that the series is Gaussian.

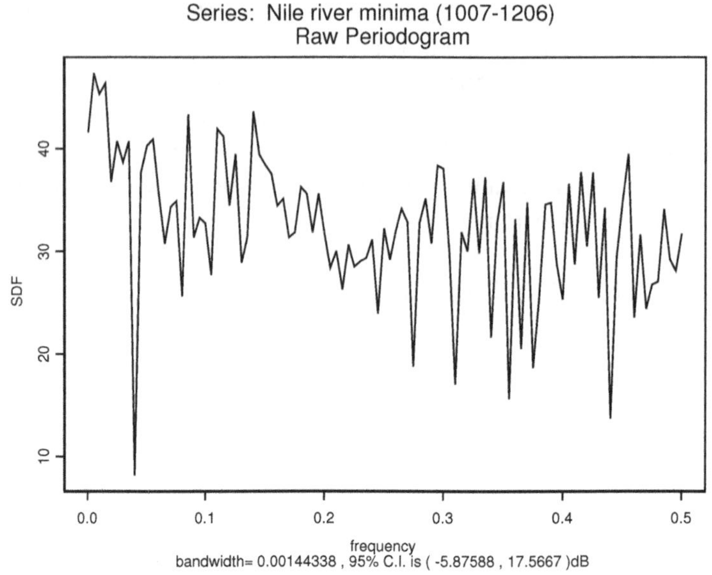

Figure 4.21: Nile river series: periodogram.

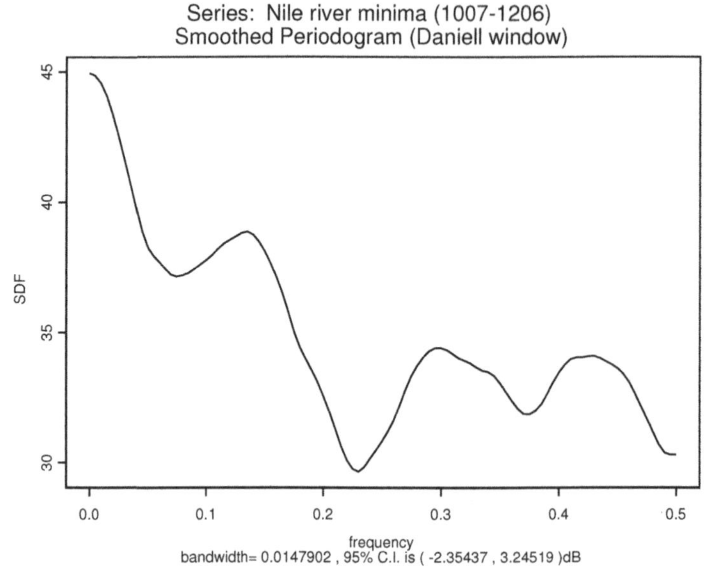

Figure 4.22: Nile river series: Daniell's window smoothed periodogram.

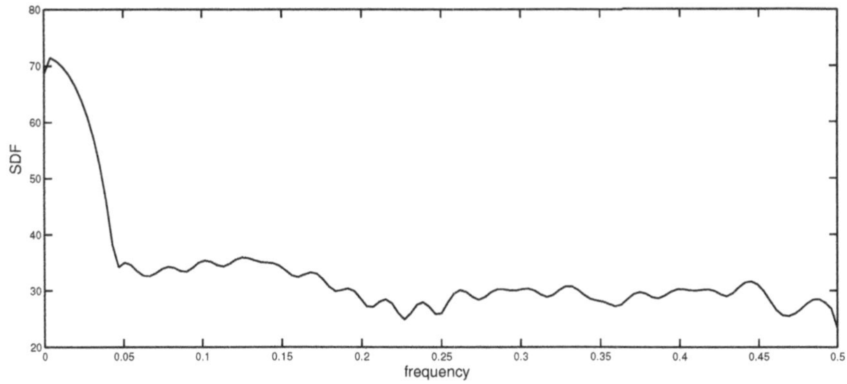

Figure 4.23: Nile river series: estimated periodogram by the WOSA method.

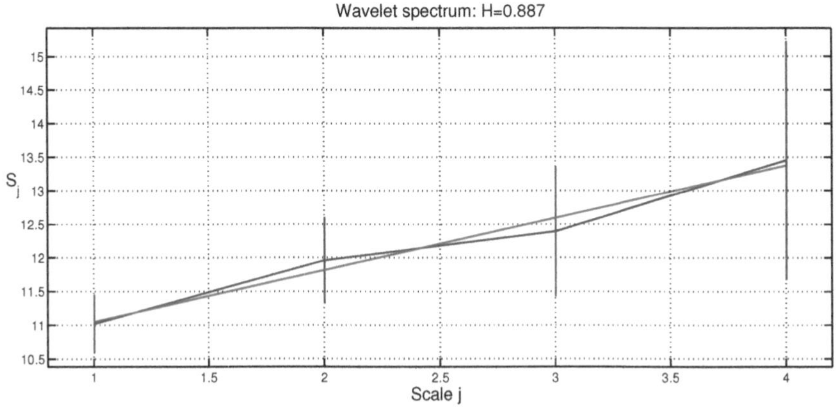

Figure 4.24: Nile river series: wavelet spectrum.

Figure 4.26 shows a graph with the theoretical SACF and ACFs of the estimated ARFIMA(0; 0.4125; 0) and AR(8) models. The AR(8) process is the best AR model according to the AIC criterium and has been estimated by the function `S-PLUS`® `AR` (Yule-Walker method). The figure shows that the AR(8) process does not model the slow decay of the series' SACF.

Figure 4.27 shows the QQ-plot of the residuals and Figure 4.28 shows the residuals' SACF graph. Both the QQ-plot and the SACF indicate that the residuals may be taken as WGN.

Figure 4.29 shows the AR(8) model's poles and zeros diagram. The high order adjusted AR model achieves modeling the ARFIMA(0;0.4125;0) long memory at the cost of positioning a pole close to the unit circle.

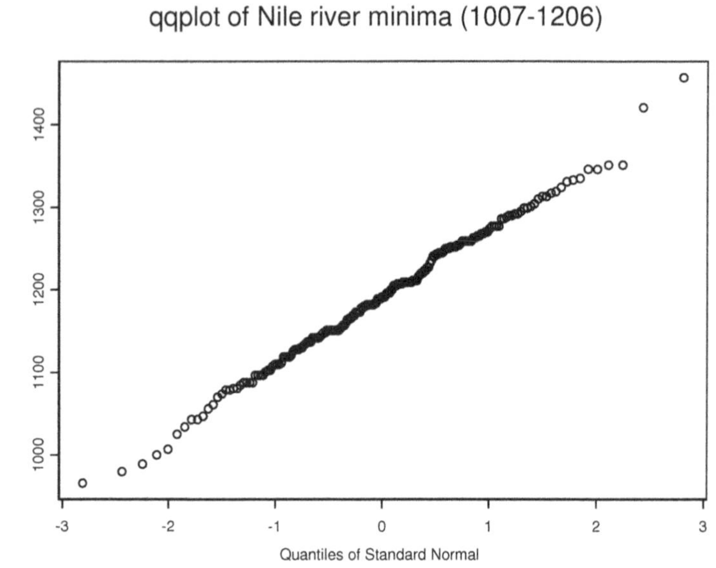

Figure 4.25: Nile river series: QQ-plot.

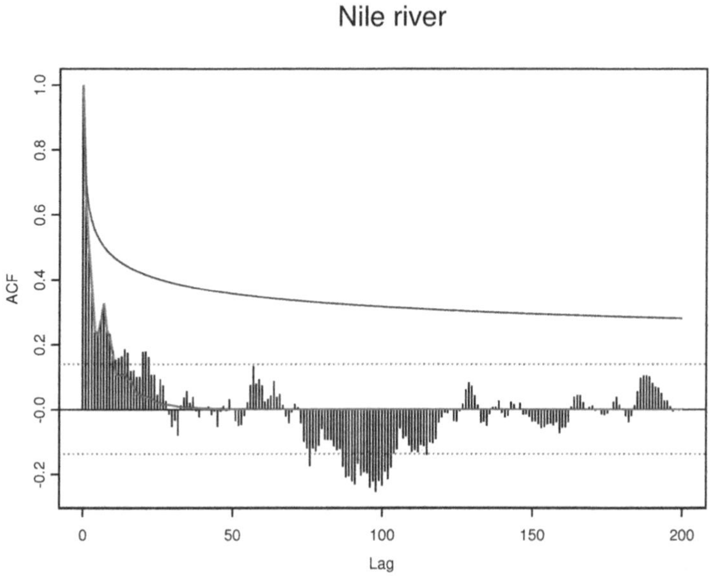

Figure 4.26: Nile river series: theoretical SACF and ACF graphs of the estimated ARFIMA(0; 0.4125; 0) (red line) and AR(8) (green line) models.

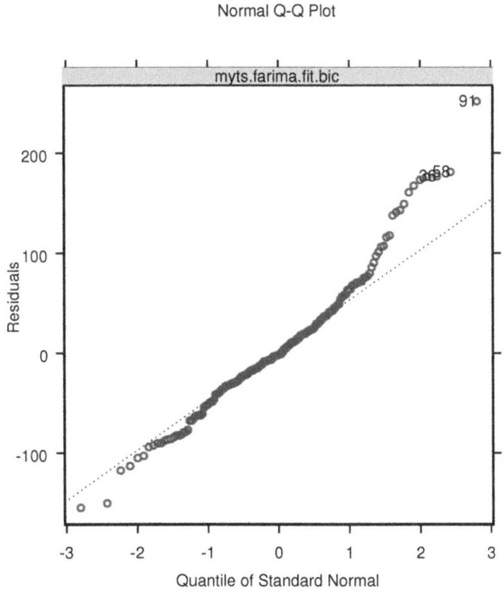

Figure 4.27: ARFIMA(0; 0.4125; 0) model's residuals QQ-plot.

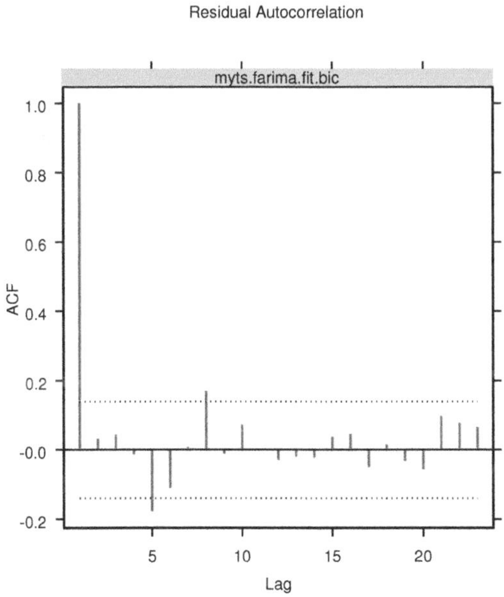

Figure 4.28: ARFIMA(0; 0.4125; 0) model's residuals SACF.

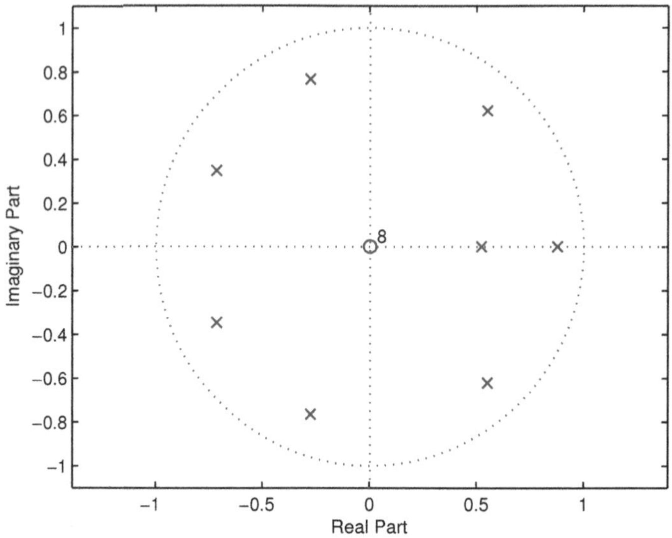

Figure 4.29: AR(8) model's poles and zeros diagram.

4.4 Prediction empirical study with the TARFIMA model

4.4.1 ARFIMA$(0, d, 0)$ series

Figures 4.30 and 4.31 illustrate the future values predictions (Type-I and Type-II) of the ARFIMA(0; 0.4; 0) with origin at $t = 3996$, horizons $h = 1, 2, \ldots, 100$ and 95% confidence intervals that were obtained by means of the TARFIMA(100) and AR(15) (Figure 4.30) and the TARFIMA(100) and AR(2) (Figure 4.31) models. The TARFIMA(100) model used the (conservative) estimate $\hat{d} = 0.3684$.

Figure 4.32 shows the difference between the absolute prediction errors of the AR(15) and TARFIMA(100) models (graph $(\Delta_h(1) - \Delta_h(2))$ *vs.* t, in which:

- $\Delta_h(1)$ is the absolute prediction error of the AR(15) model and
- $\Delta_h(2)$ is the absolute prediction error of the TARFIMA(100)) model for the ARFIMA(0; 0, 4; 0) realization.

Figure 4.33 shows the difference between the absolute prediction errors of the AR(2) and TARFIMA(100) (graph $(\Delta_h(1) - \Delta_h(2))$ *vs.* t, in which:

- $\Delta_h(1)$ is the absolute prediction error of the AR(2) model and

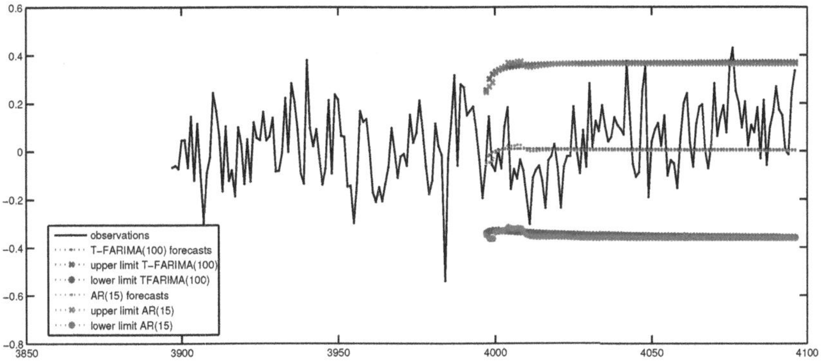

Figure 4.30: ARFIMA(0; 0.4; 0) series: observations, h-steps predictions, and 95% confidence intervals for the TARFIMA(100) and AR(15) models predictions.

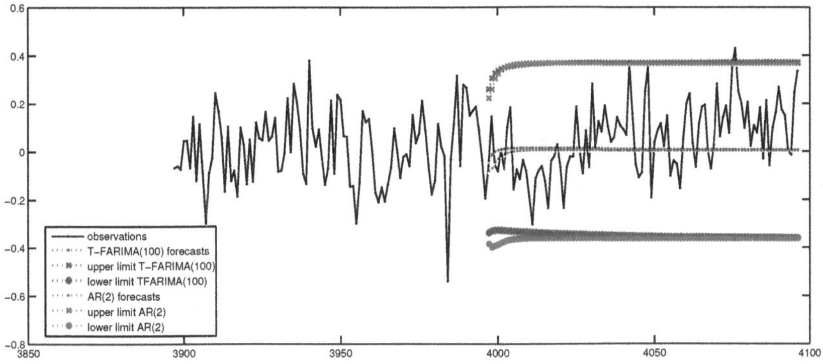

Figure 4.31: ARFIMA(0; 0.4; 0) series: observations, h-steps predictions, and 95% confidence intervals for the TARFIMA(100) and AR(2) models predictions.

- $\Delta_h(2)$ is the absolute prediction error of the TARFIMA(100)) model for the same realization.

Figures 4.32 e 4.33 suggest that the TARFIMA(100) model has a better performance because the absolute prediction errors values of the AR models have a tendency to be larger than those of the TARFIMA(100) process (note that the number of positive values of $(\Delta_h(1) - \Delta_h(2))$ in Figs. 4.32 and 4.33 is larger than the number of negative values).

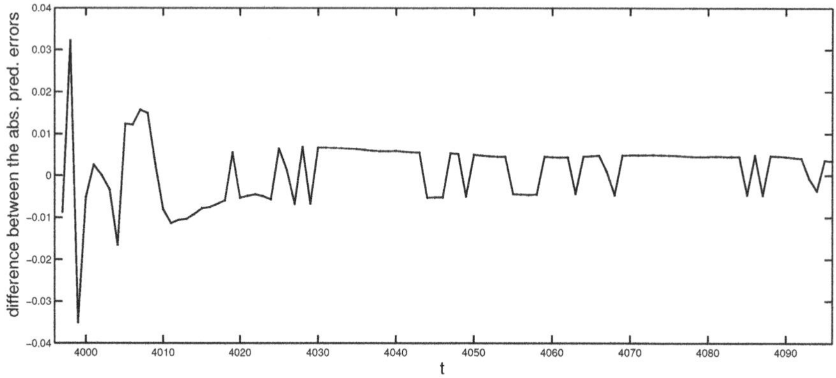

Figure 4.32: ARFIMA(0; 0.4; 0) series: TARFIMA100-AR15 - difference between the absolute prediction errors of the AR(15) and TARFIMA(100) models (graph $(\Delta_h(1) - \Delta_h(2))$ *vs.* t

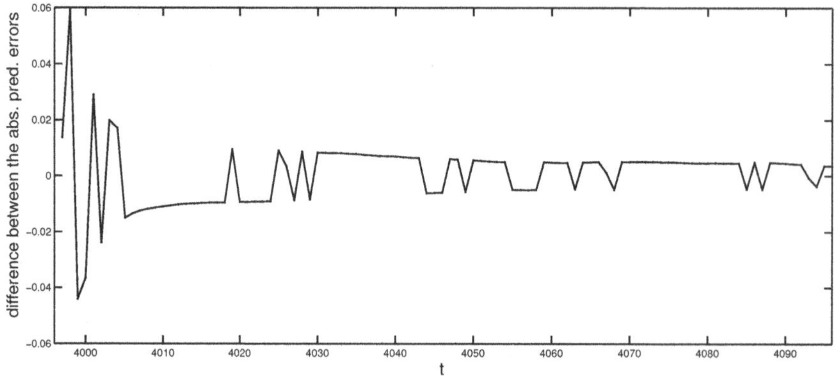

Figure 4.33: ARFIMA(0; 0.4; 0) series: TARFIMA100-AR2 - difference between the absolute prediction errors of the AR(2) and TARFIMA(100) models (graph $(\Delta_h(1) - \Delta_h(2))$ *vs.* t

It is also interesting to compare the performance of the models in terms of the *proportional cummulative empirical mean square error* (*PCEMSE*) variation [12], that will be defined in the sequence.

First, the *cummulative empirical mean square error* (*CEMSE*) with origin in t and maximum horizon K ($h = 1, 2, \ldots, h_{\max} = K$) of a model $\boldsymbol{x}_t$ is defined as [83]

$$CEMSE_{x(t,K)} = \frac{1}{K}\sum_{h=1}^{K}(x_{t+h} - \hat{x}_{t+h})^2. \tag{4.24}$$

The *prediction cummulative empirical mean square error* (*PCEMSE*) of the model $\boldsymbol{x}_t$ in relation to model $\boldsymbol{y}_t$ with origin in t and horizon K is given by

$$PCEMSE_{x,y}(t, K) = \frac{CEMSE_{x(t,K)} - CEMSE_{y(t,K)}}{CEMSE_{y(t,K)}}. \tag{4.25}$$

We say that $\boldsymbol{x}_t$ is a better (or more precise) prediction model than $\boldsymbol{y}_t$ if $PCEMSE_{x,y}(t, K) < 0$.

Table 4.9 shows that the predictions of the AR(15), TARFIMA(10), TARFIMA(50) TARFIMA(100) models for the ARFIMA(0; 0.4; 0) series are better than the AR(2) model's predictions and that the AR(15), TARFIMA(50) and TARFIMA(100) models have similar performances, according to the PCEMSE criterium.

Table 4.9: ARFIMA(0; 0.4; 0) series: $PCEMSE_{x,y}(t = 3996, K)$ of the AR(15), TARFIMA(10), TARFIMA(50) and TARFIMA(100) estimated models (columns in which $L = 10, 50,$ and 100, respectively) relative to the AR(2) model. The TARFIMA(L) models used $\hat{d} = 0.3684$.

K	AR(15)	L=10	L=50	L=100
7	-0.2807	-0.0956	-0.2839	-0.2882
10	-0.3451	-0.1529	-0.3401	-0.3438
20	-0.4203	-0.2187	-0.4068	-0.4092
50	-0.0517	-0.0262	-0.0917	-0.0840
100	0.0000	-0.0095	-0.0145	-0.0239

4.4.2 MWM series

Figures 4.34 and 4.35 illustrate the future values predictions (Type-I and Type-II) of the MWM($H = 0.9$) series with origin at $t = 3996$, horizons $h = 1, 2, \ldots, 100$ and 95% confidence intervals that were obtained by means of the TARFIMA(100) and AR(28) (Figure 4.34) and the TARFIMA(100) and AR(2) (Figure 4.35) models. The TARFIMA(100) model used the (conservative) estimate $\hat{d} = 0.3572$.

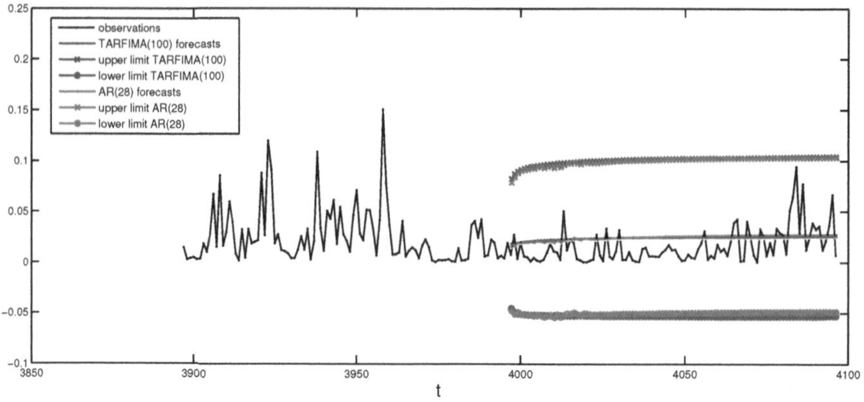

Figure 4.34: MWM($H = 0.9$) series: h-steps predictions, and 95% confidence intervals for the TARFIMA(100) and AR(28) models predictions *vs.* t

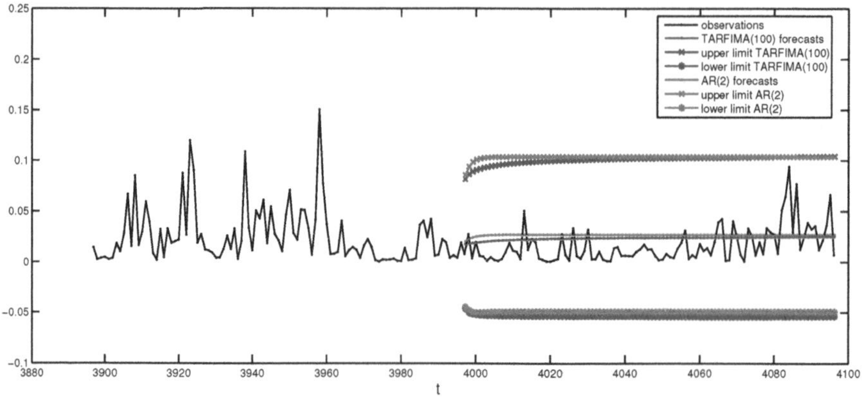

Figure 4.35: MWM($H = 0.9$) series: h-steps predictions, and 95% confidence intervals for the TARFIMA(100) and AR(2) models predictions *vs.* t

Figure 4.36 shows the difference between the absolute prediction errors of the AR(28) and TARFIMA(100) models (graph $(\Delta_h(1) - \Delta_h(2))$ *vs.* t, in which:

- $\Delta_h(1)$ is the absolute prediction error of the AR(28) model and
- $\Delta_h(2)$ is the absolute prediction error of the TARFIMA(100)) model for the MWM($H = 0.9$) series realization.

Figure 4.37 shows the difference between the absolute prediction errors of the AR(2) and TARFIMA(100) (graph $(\Delta_h(1) - \Delta_h(2))$ *vs.* t, in which:

- $\Delta_h(1)$ is the absolute prediction error of the AR(2) model and
- $\Delta_h(2)$ is the absolute prediction error of the TARFIMA(100)) model for the same realization.

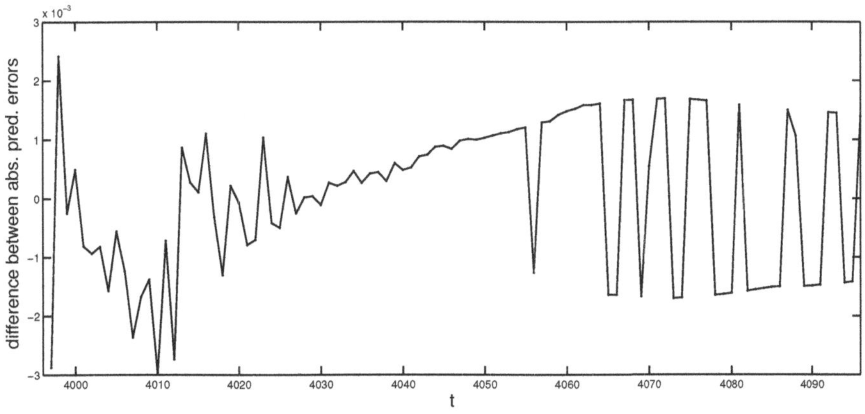

Figure 4.36: MWM($H = 0.9$) series: difference between of prediction errors absolute values of the AR(28) and TARFIMA(100) models

Table 4.10 ($PCEMSE_{x,y}(t = 3996, K)$ for the MWM series with $H = 0.9$) shows that the AR(28), TARFIMA(10), TARFIMA(50) and TARFIMA(100) models present better performance than the AR(2) model for $K = \{50, 100\}$, according to the PCEMSE criterium. Also note that the TARFIMA(50) and TARFIMA(100) models are slightly better than the AR(2) for $K = 7$. The AR(28), TARFIMA(50) and TARFIMA(100) have similar performances for $K = 100$.

4.4.3 Nile river series between years 1007 and 1206

Figures 4.38 and 4.39 illustrate the future values predictions (Type-I and Type-II) of the Nile river series with origin at $t = 1106$, horizons $h = 1, 2, \ldots, 100$ and 95% confidence intervals that were obtained by means of the TARFIMA(100) and AR(8) (Figure 4.38) and the TARFIMA(100)

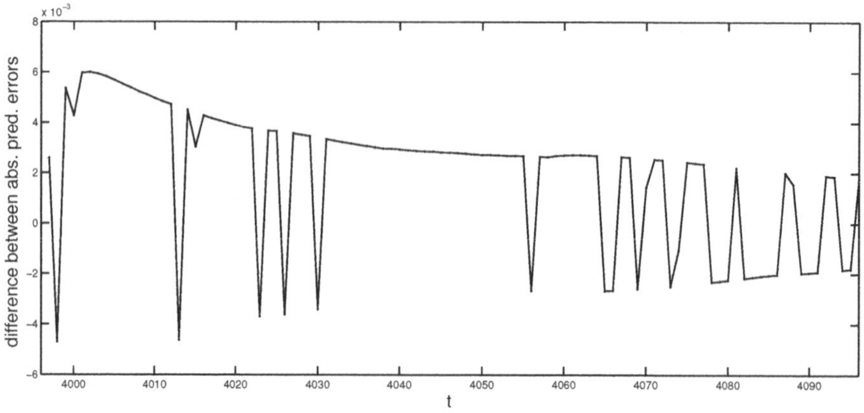

Figure 4.37: MWM($H = 0.9$) series: difference between of prediction errors absolute values of the AR(2) and TARFIMA(100) models

Table 4.10: MWM($H = 0.9$) series: $PCEMSE_{x,y}(t = 3996, K)$ of the estimated AR(28), TARFIMA(10), TARFIMA(50) and TARFIMA(100) models relative to the AR(2). The TARFIMA(L) models used $\hat{d} = 0.3572$.

K	AR(28)	L=10	L=50	L=100
7	0.0858	0.0118	-0.0148	-0.0148
10	0.1119	0.1891	0.0871	0.0473
20	0.1863	0.0668	0.2205	0.2888
50	-0.0865	-0.0673	-0.0625	-0.0024
100	-0.1671	-0.0430	-0.1797	-0.1646

and AR(2) (Figure 4.39) models. The TARFIMA(100) model used the (conservative) estimate $\hat{d} = 0.4125$.

Figure 4.40 shows the difference between the absolute prediction errors of the AR(28) and TARFIMA(100) models (graph $(\Delta_h(1) - \Delta_h(2))$ *vs.* t, in which:

- $\Delta_h(1)$ is the absolute prediction error of the AR(8) model and
- $\Delta_h(2)$ is the absolute prediction error of the TARFIMA(100)) model for the Nile river series realization.

Figure 4.41 shows the difference between the absolute prediction errors of the AR(2) and TARFIMA(100) (graph $(\Delta_h(1) - \Delta_h(2))$ *vs.* t, in which:

- $\Delta_h(1)$ is the absolute prediction error of the AR(2) model and

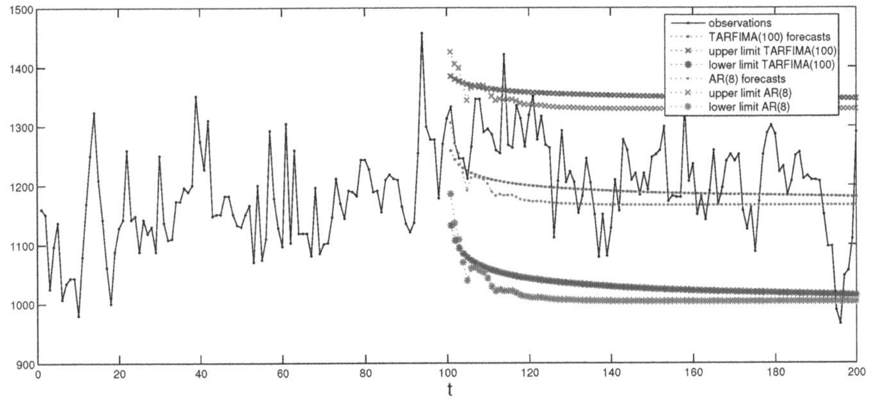

Figure 4.38: Nile river series: Série do rio Nilo: h-steps predictions, and 95% confidence intervals for the TARFIMA(100) and AR(8) models predictions *vs.* t

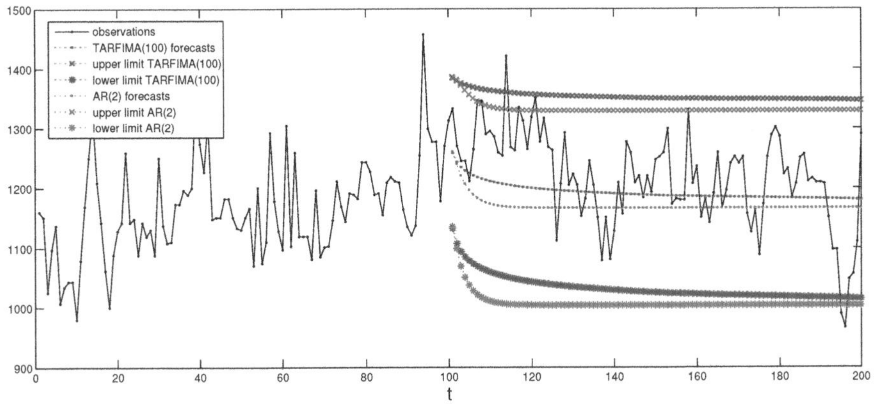

Figure 4.39: Nile river series: h-steps predictions, and 95% confidence intervals for the TARFIMA(100) and AR(2) models predictions *vs.* t

- $\Delta_h(2)$ is the absolute prediction error of the TARFIMA(100)) model for the same series.

Table 4.11 ($PCEMSE_{x,y}(t = 1106, K)$ for the Nile river series) shows that the AR(8), TARFIMA(10), TARFIMA(50) and TARFIMA(100) models present better performance than the AR(2) model for $K = \{100\}$, according to the PCEMSE criterium. Also note that the TARFIMA(50) and

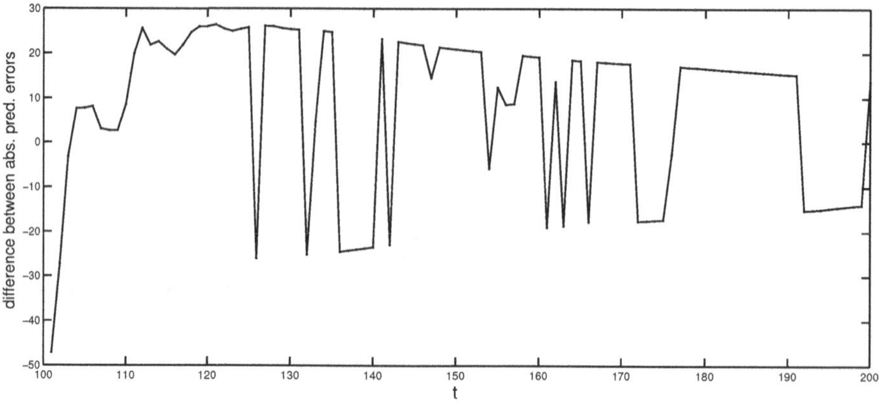

Figure 4.40: Nile river series: difference between of prediction errors absolute values of the AR(8) and TARFIMA(100) models

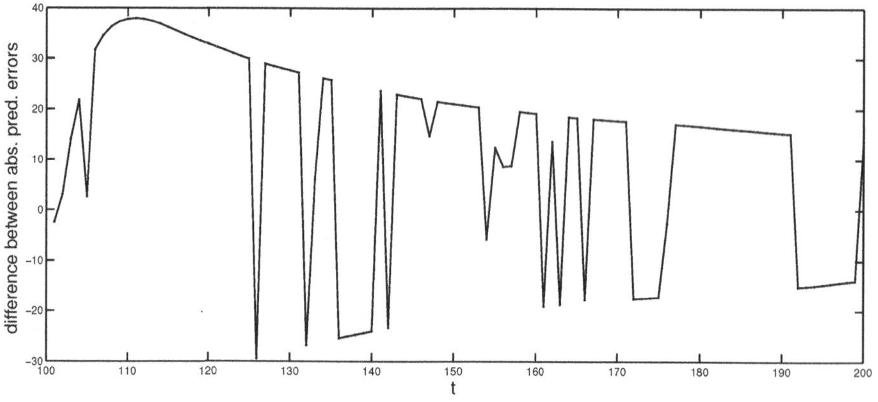

Figure 4.41: Nile river series: difference between of prediction errors absolute values of the AR(2) and TARFIMA(100) models

TARFIMA(100) have similar performances for $K = 100$ and are much better than the AR(8).

4.4.4 Conclusions

The study presented in this Section validates the application of the long memory TARFIMA(L) model in (Gaussian/non-Gaussian) LRD signals.

The simulations suggest that long memory signals modeling by means of AR processes is feasible, at least from a theoretical point of view. The

Table 4.11: Nile river series: $PCEMSE_{x,y}(t = 1106, K)$ of the estimated AR(8), TARFIMA(10), TARFIMA(50) and TARFIMA(100) models relative to the AR(2). The TARFIMA(L) models used $\hat{d} = 0.4125$.

K	AR(8)	L=10	L=50	L=100
7	0.0796	0.1331	0.1852	0.2128
10	0.0925	0.0649	0.1332	0.1687
20	0.0516	0.0871	0.1864	0.2570
50	-0.0145	-0.0061	0.0098	0.0559
100	-0.1024	-0.1033	-0.2602	-0.2617

computer experiments with ARFIMA and MWM series show that high order AR models may offer a competitive prediction performance. The statistical inference procedure "works" because the number N of signal points is sufficiently large (i. e., the fact that the estimated AR model is not parsimonious is not a problem).

However, note that some relevant questions from the practical point of view arise:

- the long memory modeling may be obtained at the cost of placing one or more poles close to the unit circle; therefore the implementation of the fitted AR model may be unstable due to quantization effects of the digital filter coefficients.
- the model's *order selection fundamental problem.* Note that selection criteria as the AIC don't take into account, explicitly, the $1/f^{\alpha}$-type singularity that exists at the SDF origin of LRD series. Consider the statements made by Percival and Walden [96, p.434]:

  > "*First, any order selection method we use should be appropriate for what we intend to do with the fitted AR model.* (...) *Most of the commonly used order selection criteria are geared toward selecting a low-order AR model that does well for one-step-ahead predictions*".

  Following, Percival and Walden describe four selection methods (including the AIC). At last, below there is a citation of two important remarks [96, p.437]:

  > "*How well do these order selection criteria work in practice?* (...) *They conclude that subjective judgment is still needed to select the AR order for actual time series.*".

We realize that the application of AR models in on-line prediction schemes is strongly impaired by model's order identification problem. On the other hand, the TARFIMA(L) process has the following advantages:

1. it models the spectrum's low frequencies region of LRD signals;
2. it may implemented in practical applications as CAC and bandwidth dynamic allocation, because it is finite-dimensional (it does not require an infinite memory as the ARFIMA model does);
3. the state space representation enables the use of the Kalman filter.

5

Modeling of Internet traffic

This chapter uses the concepts developed in the previous chapters to model the Internet traffic and to produce accurate forecasts.

5.1 Introduction

This chapter presents a case study based on the analysis of the `2002_Apr_13_Sat_1930.7260.sk1.1ms.B_P.ts.gz` Internet traffic trace [91, 92]. This series, that is part of a set of 20 traces called "UNC02", contains 2 hours collection of unidirectional IP traffic at the 1 milisecond scale of a Gigabit Ethernet link between the University of North Carolina (UNC) campus at Chapel Hill and its ISP on 13/04/2002 from 19:30 thru 21:30. This trace, that consists of the register of packets and bytes count time series in the inbound direction(i. e., the traffic entering the UNC's campus) during 1 milisecond bin, has been collected by the `tcpdump` program, originally developed by Van Jacobson, Leres and McCanne [56], and is available at `http://www-dirt.cs.unc.edu/ts/`. According to authors of the study [92], the collection has been implemented with a 1 microsecond precision, and a 0.03% packets loss rate has been observed for this trace. Therefore, the effect of this inaccuracy is negligible with 1 or 10 ms bins.

5.2 Modeling of the UNC02 trace

5.2.1 Exploratory analysis

In this section, we make an exploratory analysis of the packets and bytes times series of the `UNC02` at the 1 millisecond scale and at the 1 second aggregated scale.

One of the most important objectives of time series analysis is to build a stochastic model of the series under investigation. The model can be stationary or non-stationary.

ARMA or ARFIMA models, for example, assume that the series is stationary. However, most of the empirical series may present some form of non-stationarity. A series may be stationary for a very long time or may be stationary during very short intervals, changing its level or trend. This being so, it is important to try to answer the following preliminary questions before proceeding with a more detailed exploratory analysis [66, 122]:

- Is the series under investigation globally stationary?
- If the series is not globally stationary, is it possible to identify a chain of stationary sub-series, that can be generated from stationary models?
- Does the series present deterministic and/or stochastic tendencies, SRD e/ou LRD?
- How can we detect or estimate regime changes (known in the literature as the change-point problem) of a non-stationary series?

The *non-stationarity* of network traffic for the order of hours time scale is intuitive. Consider, for example, the use of an Internet access link in a corporate network as the University of São Paulo - USP's. There is no doubt that the use of this link during the morning and afternoon peak hours should be higher than during the dawn period. It is also expected that this traffic presents some kind of deterministic trend like a "raise gradient" just after the beginning of the working hours, around 8 o'clock in the morning. So, it makes sense to state that the networks traffic is non-stationary as for the mean and/or trend at the hours scale.

On the other hand, several studies [69, 92, 94] show that it is reasonable to characterize the Internet traffic as self-similar and LRD (remember that the long memory concept has been defined only for wide sense stationary processes) at scales ranges that may go from hundreds of milliseconds to hundreds of seconds (the low and high limits may vary according to the specific network).

Figure 5.1 shows the inbound traffic (packets per bin count series) on an access Gigabit Ethernet link of UNC on 13/04/2002, time 19:30-21:30, at the 1 millisecond scale. Figure 5.2 shows the inbound traffic (packets per bin count series) on an access Gigabit Ethernet link of UNC on 13/04/2002, time 19:30-21:30, at the 1 second scale. In both scales it is possible to observe that the traffic is highly impulsive which is already an indication of LRD.

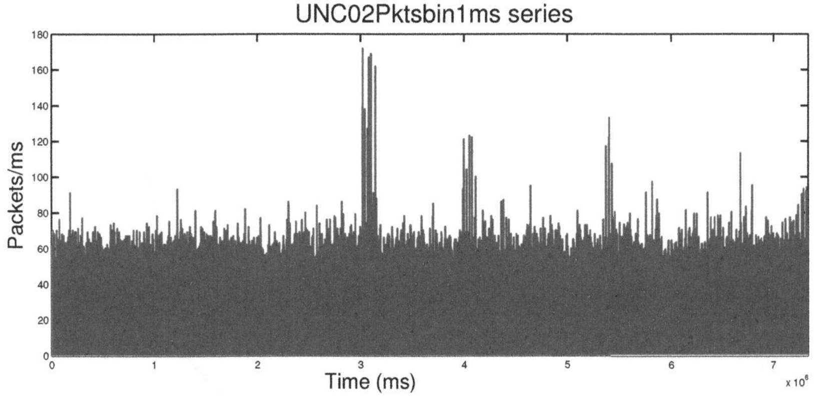

Figure 5.1: Inbound traffic (packets per bin count series) on an access Gigabit Ethernet link of UNC on 13/04/2002, time 19:30-21:30, 1 millisecond scale.

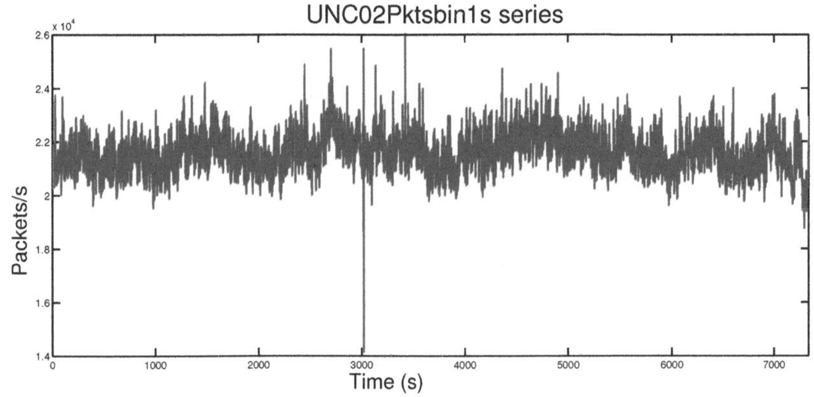

Figure 5.2: Inbound traffic (packets per bin count series) on an access Gigabit Ethernet link of UNC on 13/04/2002, time 19:30-21:30, 1 second scale.

Figure 5.3 shows the QQ-plot of the `UNC02Pktsbin1ms`. The figure indicates that the series deviates strongly from normality.

Figure 5.4 shows the QQ-plot of the `UNC02Pktsbin1s`. In this case, it is difficult to arrive at any conclusion about normality and it is necessary to employ formal statistical tests.

The Jarque-Bera's normality test [58] (based on asymmetry and kurtosis measurements) rejected the null hypothesis of normality of

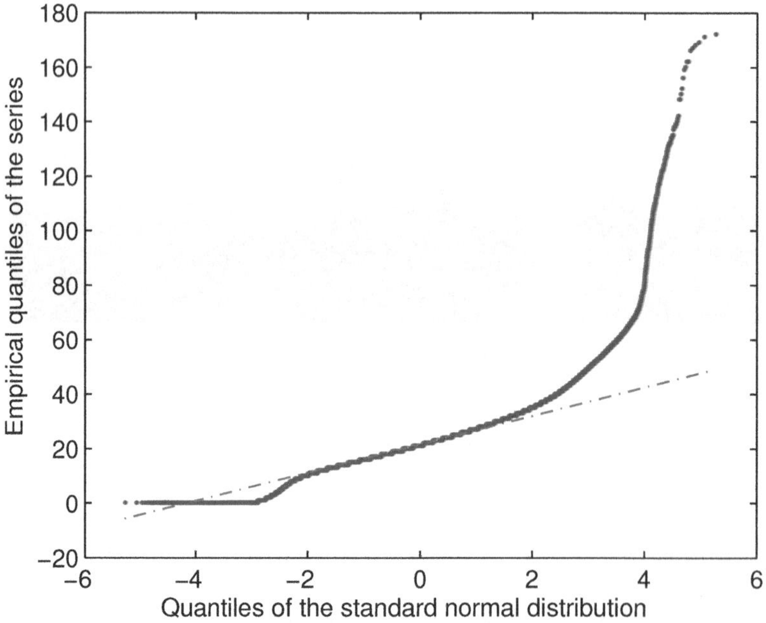

Figure 5.3: QQ-plot of `UNC02Pktsbin1ms`.

`UNC02Pktsbin1s`. The statistics of this test is defined as

$$JB = \frac{N}{6}\left(\hat{A}^2 + \frac{(\hat{K}-3)^2}{4}\right), \tag{5.1}$$

in which N is the number of samples, $\hat{A}$ is the asymmetry estimator

$$\hat{A} = \frac{1}{Ns_x^3}\sum_{t=1}^{N}(x_t - \hat{\mu})^3 \tag{5.2}$$

in which $\hat{\mu}$ denotes the sample mean, s_x^2 is the sample variance and $\hat{K}$ is the kurtosis estimator

$$\hat{K} = \frac{1}{Ns_x^4}\sum_{t=1}^{N}(x_t - \hat{\mu})^4. \tag{5.3}$$

Under the null hypothesis that the data are normally distributed, we expect $JB \sim \chi^2(2)$ (chi-square with two degrees of freedom).

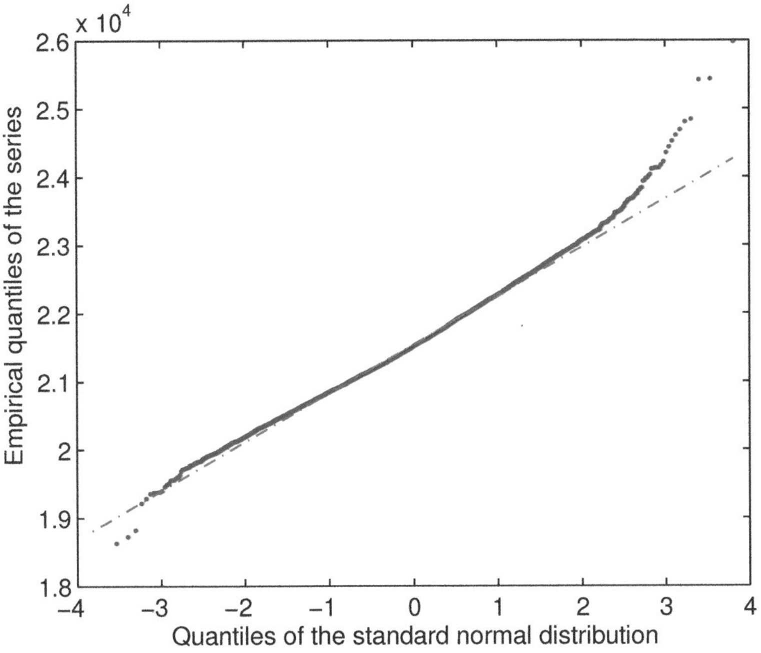

Figure 5.4: QQ-plot of `UNC02Pktsbin1s`.

The program `STABLE4.0` for `MATLAB`, provided by Dr. John P. Nolan (Robust Analysis, Inc.) to be used in this research, fitted a stable distribution $S_\alpha(\sigma, \eta, \mu) = S_{1.92}(495; 0.86; 2.15 \times 10^4)$ to `UNC02Pktsbin1s`.

Figure 5.5 illustrates the graphs produced by the `stablediag` function, namely:

- PDF fitting,
- QQ-plot,
- PP(*Probability-Probability*)-plot and
- ZZ-plot.

The PP-plot is built using the theoretical CDF, $F_x(x)$, of the fitted model. The data sample values, sorted in increasing order, are denoted by $x(1), x(2), \ldots, x(n)$. For $i = 1, 2, \ldots, n$, $F_x(x(i))$ (theoretical value) is represented *versus* $q_i = (i - 0, 5)/n$ (CDF empirical value).

The ZZ-plot is the inverse graph of the theoretical CDF *vs* the inverse of the empirical CDF (i. e., it may be obtained from the PP-plot).

According to Nolan, the diagnostic of stable data by means of the QQ-plot presents the following practical problems:

- a) most of the data stays compressed in a small region and
- b) there is a significative data deviation at the ends of the QQ-plot relative to the theoretical line.

For this reason Nolan recommends using the PP-plot and ZZ-plot as diagnostic tools.

So, the `UNC02Pktsbin1s` series has heavy tail and is well modeled by a stable distribution, according to the `STABLE4.0` toolbox.

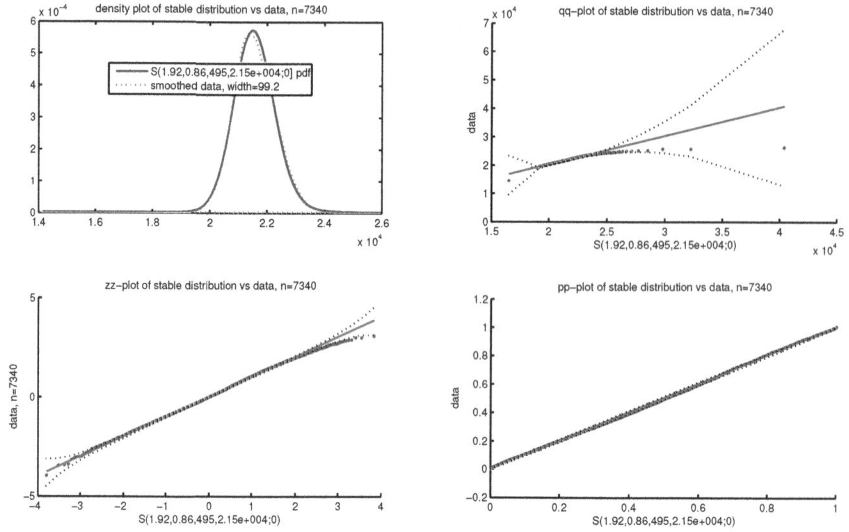

Figure 5.5: Fitting of a stable distribution with parameters $\{\alpha = 1.92, \sigma = 495, \eta = 0.86, \mu = 2.15 \times 10^4\}$, to the `UNC02Pktsbin1s` series .

Figure 5.6 shows the wavelet spectrum of the `UNC02` trace (in packets/*bin*) estimated by means of the `UNC02Pktsbin1ms` series. At this scale it is possible to observe that the traffic presents both LRD and SRD. LRD is manifest at higher values of the scale parameter J (lower frequencies), while SRD manifests at lower values of J (higher frequencies).

Figure 5.7 shows the wavelet spectrum of the `UNC02` trace (in packets/*bin*) estimated by means of the `UNC02Pktsbin1s` series. At this scale only the LRD behavior is visible. This means that the aggregation has smoothed out the SRD behavior.

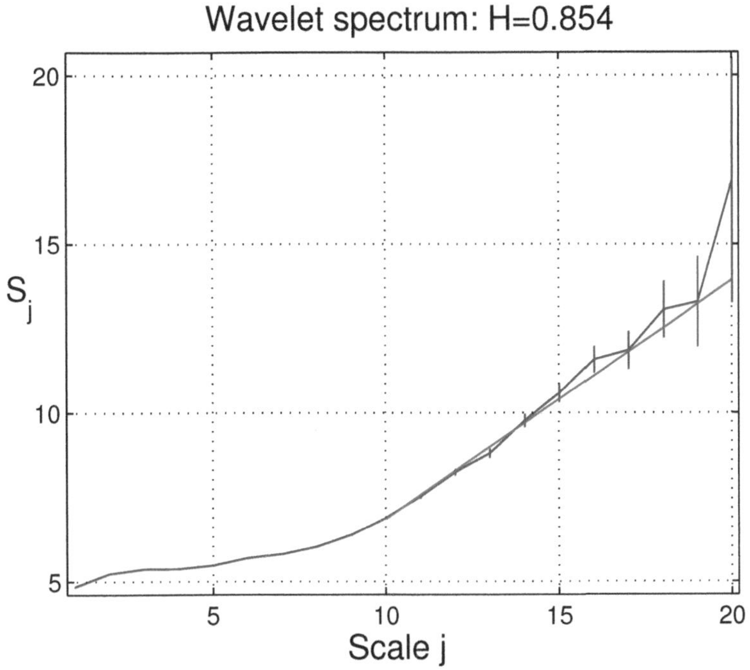

Figure 5.6: Wavelet spectrum of the `UNC02` trace (in packets/*bin*) estimated by means of the `UNC02Pktsbin1ms` series.

Figure 5.8 shows the inbound traffic (bytes per bin count series) on an access Gigabit Ethernet link of UNC on 13/04/2002, time 19:30-21:30, at the 1 millisecond scale. The average traffic is 51.623 Mbps (or 6.4529 Mbytes/s), implying a link utilization of approximately 5, 16%.

Figure 5.9 shows the inbound traffic (bytes per bin count series) on an access Gigabit Ethernet link of UNC on 13/04/2002, time 19:30-21:30, at the 1 second scale. The average traffic is 51.623 Mbps (or 6.4529 Mbytes/s), implying a link utilization of approximately 5, 16%.

In both figures, 5.8 and 5.9, impulsiveness is clearly present.

Table 5.1 summarizes the statistical tests indicating that the trace is stationary, presents LRD, is not normal, is linear and does not have unit roots. The Hinich tests, that use the FFT (*Fast Fourier Transform*), were applied to the first block of 4096 samples of `UNC02bin1s` (that has 7340 samples, a number that is not a power of 2).

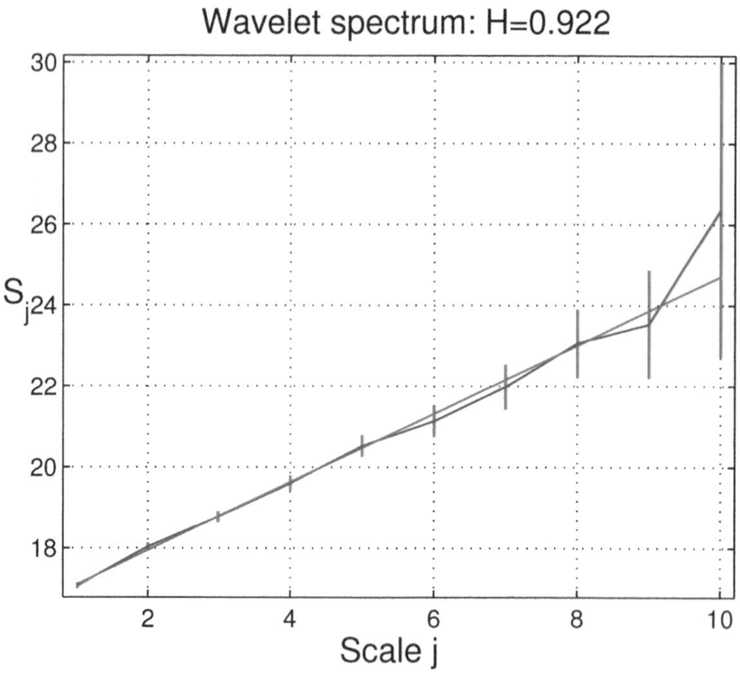

Figure 5.7: Wavelet spectrum of the `UNC02` trace (in packets/*bin*) estimated by means of the `UNC02Pktsbin1s` series.

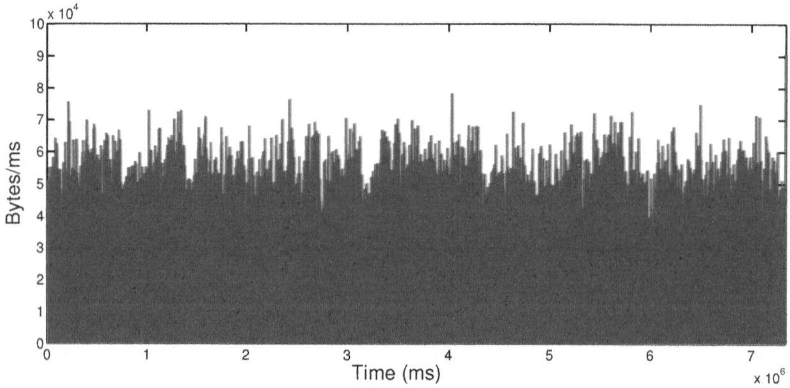

Figure 5.8: Inbound traffic (bytes per bin count series) on an access Gigabit Ethernet link of UNC on 13/04/2002, time 19:30-21:30, 1 millisecond scale.

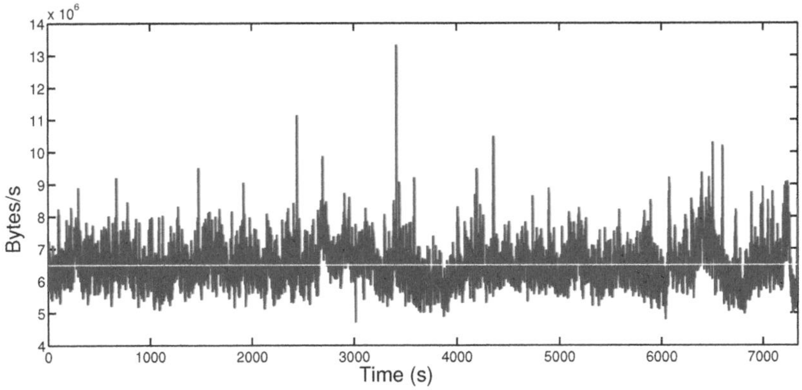

Figure 5.9: Inbound traffic (bytes per bin count series) on an access Gigabit Ethernet link of UNC on 13/04/2002, time 19:30-21:30, 1 second scale.

Table 5.1: UNC02bin1s series: results of statistical tests (stationarity, long memory, normality and linearity) and quantification of the number of unit roots (column $\hat{m}$).

x_t is $I(0)$?	x_t is LRD?		Hinich		$\hat{m}$
KPSS	R/S	GPH	x_t is normal?	x_t is linear?	
YES	YES	YES	NO	YES	0
$p_{af} > 0.05$	$p_{af} < 0.01$	$p_{af} < 0.01$	$p_{af} = 0$		

Figure 5.10 shows the result of fitting of a stable PDF with parameters $\{\alpha = 1.79, \sigma = 4.09 \times 10^5, \eta = 0.99, \mu = 6.35 \times 10^6\}$, to the UNC02bin1s series corroborating that the trace has a heavy tail distribution.

Figure 5.11 shows the wavelet spectrum of the UNC02 trace (in bytes/*bin*) estimated by means of the UNC02Bytesbin1ms series. At this scale it is possible to observe that the traffic presents both LRD and SRD. LRD is manifest at higher values of the scale parameter J (lower frequencies), while SRD manifests at lower values of J (higher frequencies).

Figure 5.12 shows the wavelet spectrum of the UNC02 trace (in bytes/*bin*) estimated by means of the UNC02Bytesbin1s series. At this scale only the LRD behavior is visible. This means that the aggregation has smoothed out the SRD behavior.

Table 5.2 shows the d parameter estimates for the UNC02bin1s series. All estimators agree reasonably well with the exception of the R/S.

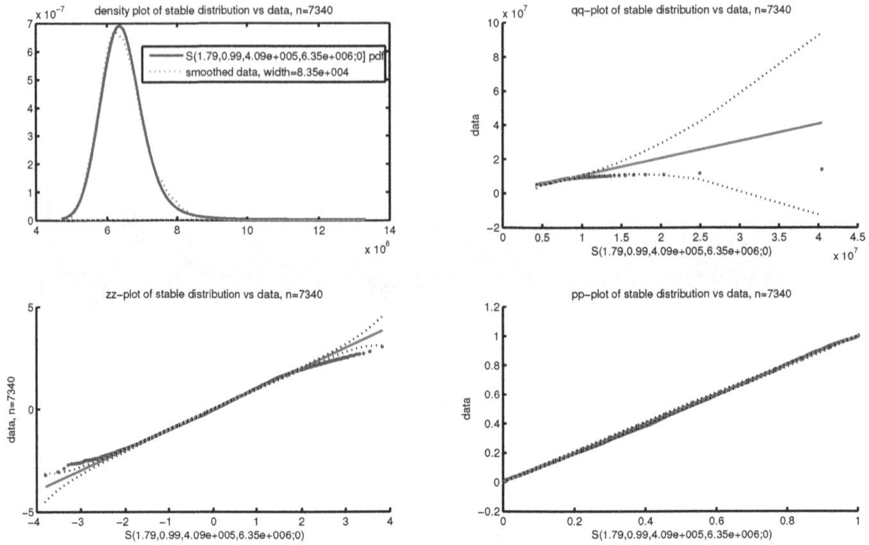

Figure 5.10: Fitting of a stable PDF with parameters $\{\alpha = 1.79, \sigma = 4.09 \times 10^5, \eta = 0.99, \mu = 6.35 \times 10^6\}$, to the `UNC02bin1s` series .

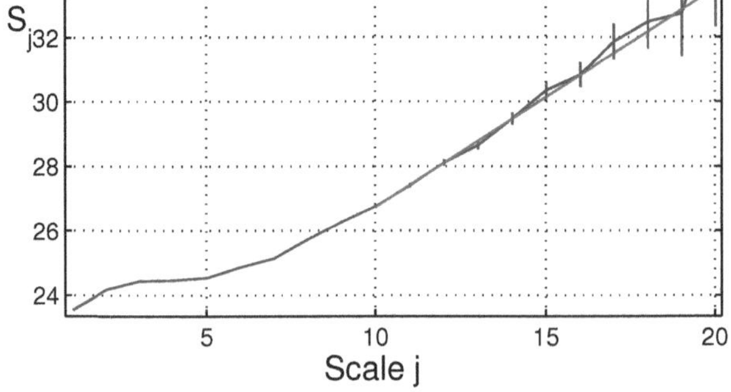

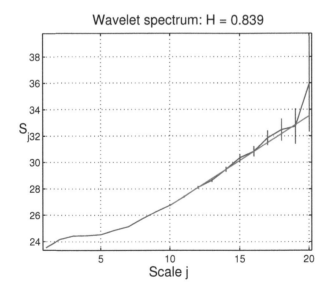

Figure 5.11: Wavelet spectrum of the `UNC02` trace (in bytes/*bin*) estimated by means of the `UNC02Bytesbin1ms` series.

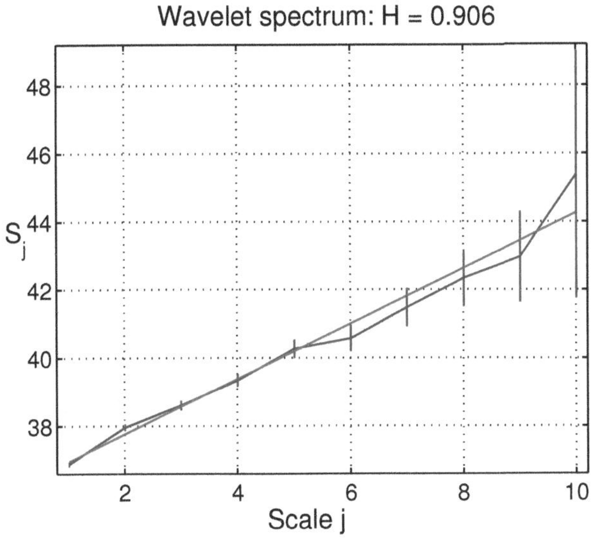

Figure 5.12: Wavelet spectrum of the `UNC02` trace (in bytes/*bin*) estimated by means of the `UNC02Bytesbin1s` series.

Table 5.2: *d* parameter estimates for the `UNC02bin1s` series.

R/S analysis	periodogram with Daniell's window	Whittle estimator	Haslett-Raftery MV estimator	Abry/Veitch wavelet estimator
0.3125	0.3854	0.4225	0.3717	0.406

Figure 5.13 shows the `UNC02bin1s` series' Daniell's window smoothed periodogram and Figure 5.14 shows the estimated periodogram by the WOSA method. In both cases it is possible to realize the increase of the SDF when $f \to 0$ indicating LRD behavior.

Figure 5.15 shows the theoretical SACF and ACFs graphs of the estimated AR(18) and ARFIMA(0; 0.3717; 1) models. It is clearly seen that the AR(18) model is not able to capture the slow decay of the SACF that is characteristic of LRD behavior. On the hand, the ARFIMA(0; 0.3717; 1) model produces a nice adjustment to the SACF slow decay.

In the sequence an analysis of residuals is presented.

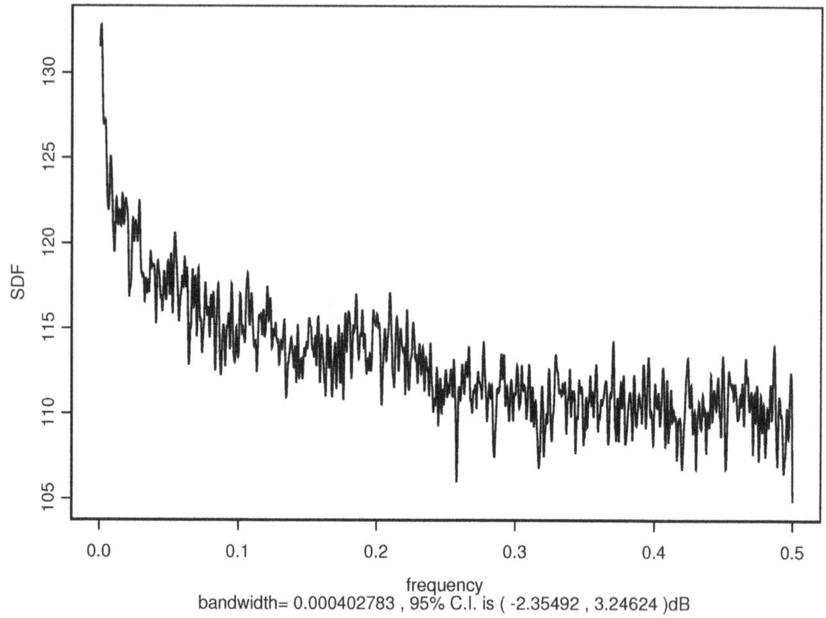

Figure 5.13: `UNC02bin1s` series: Daniell's window smoothed periodogram.

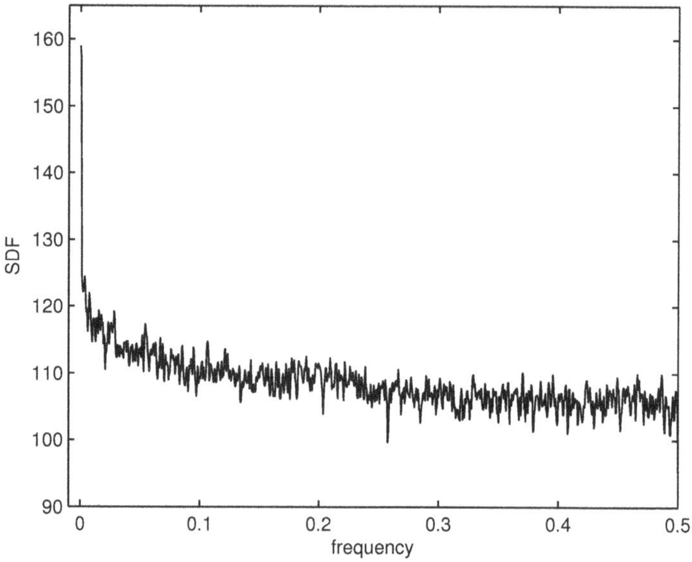

Figure 5.14: `UNC02bin1s` series: estimated periodogram by the WOSA method.

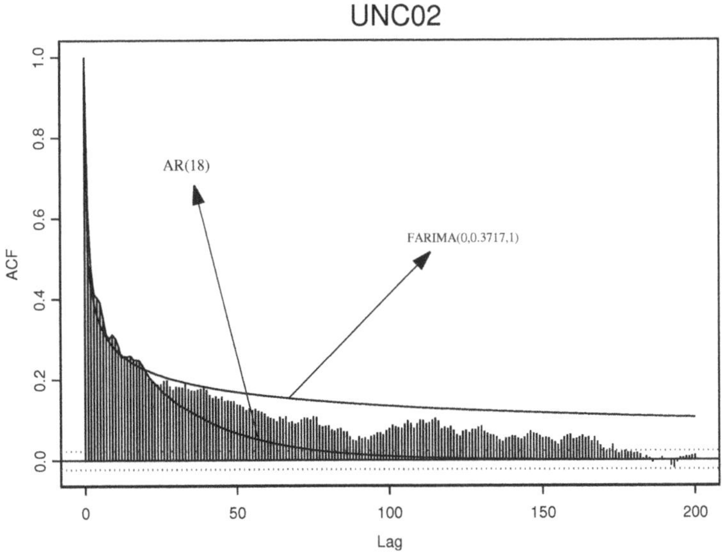

Figure 5.15: Série `UNC02bin1s`: theoretical SACF and ACFs graphs of the estimated AR(18) and ARFIMA(0; 0.3717; 1) models.

Figure 5.16 shows the QQ-plot of the residuals of the estimated ARFIMA(0; 0.3717; 1) model for the `UNC02bin1s` series.

Figure 5.17 shows the SACF of the residuals of the estimated ARFIMA(0; 0.3717; 1) model for the `UNC02bin1s` series, and Figure 5.18 shows the corresponding periodogram .

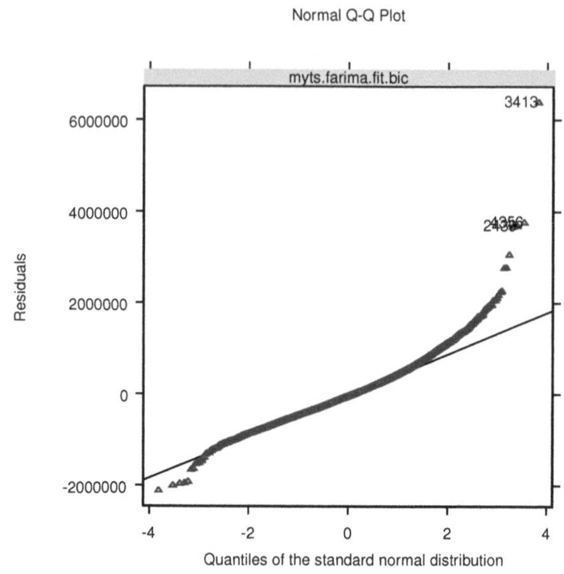

Figure 5.16: Residuals of the estimated ARFIMA(0; 0.3717; 1) model for the `UNC02bin1s` series: QQ-plot.

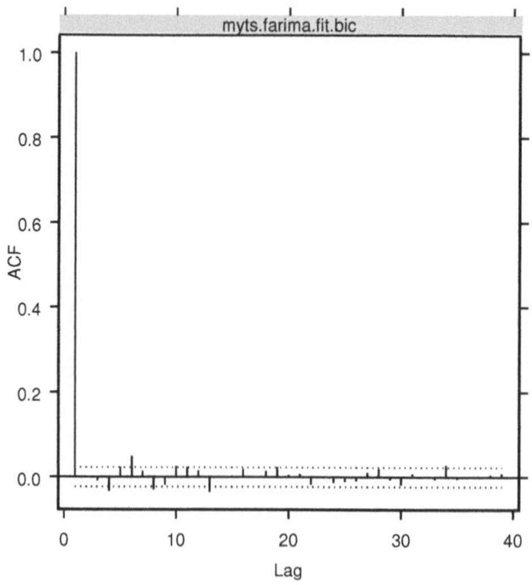

Figure 5.17: Residuals of the estimated ARFIMA(0; 0.3717; 1) model for the `UNC02bin1s` series: SACF.

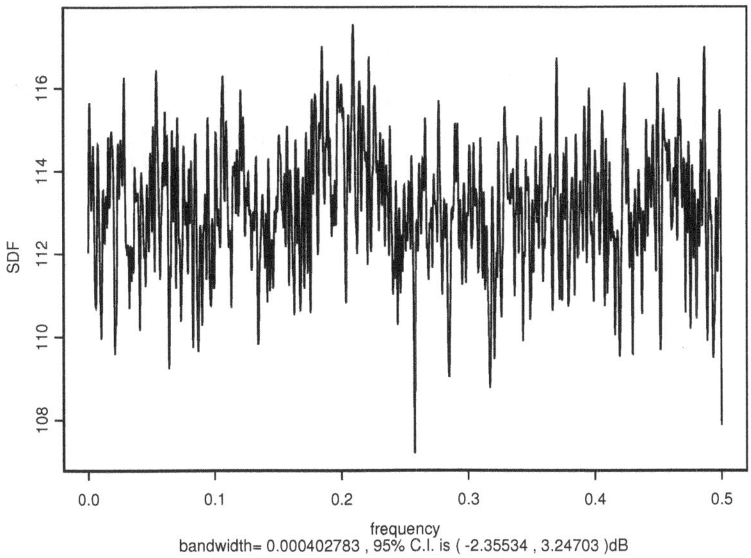

Figure 5.18: Residuals of the estimated ARFIMA(0; 0.3717; 1) model for the `UNC02bin1s` series: periodogram.

From Figures 5.16, 5.18 and 5.17 it is clear seen that the residuals marginal distribution is quite close to a Gaussian distribution, their SDF is practically flat without showing peaks, low or high-pass characteristics and the SACF presents a value different from zero only for lag equal to zero. In other words, the residuals of the estimated ARFIMA(0; 0.3717; 1) model for the `UNC02bin1s` series behaves like a WGN and, therefore, the model is a good one.

Figure 5.19 shows the periodogram of the residuals of the fitted TARFIMA(100) model for the `UNC02bin1s` series and Figure 5.20 shows the corresponding SACF.

The periodogram is practically flat and the SACF is different from zero only for the lag equal to zero. These are characteristics of a WGN indicating that the fitted TARFIMA(100) is a good model for the `UNC02bin1s` series. It is important to realize, as seen in the previous chapter, that $L = 100$ of the TARFIMA model represents the number of terms that are retained in the expansion of the ARFIMA model. The number of parameters is only two and the model is parsimonious.

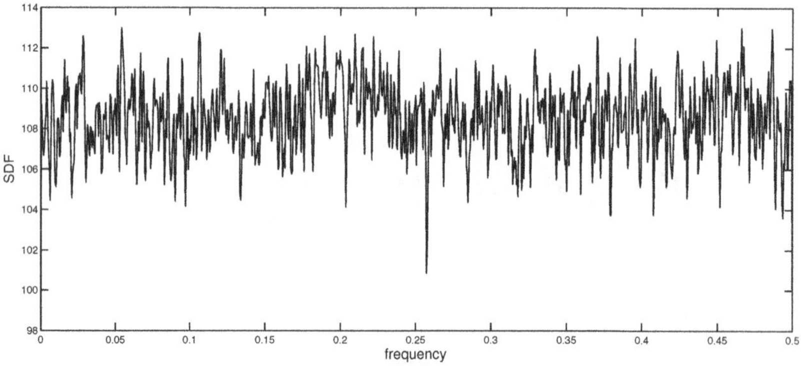

Figure 5.19: Residuals of the fitted TARFIMA(100) model for the `UNC02bin1s` series: periodogram.

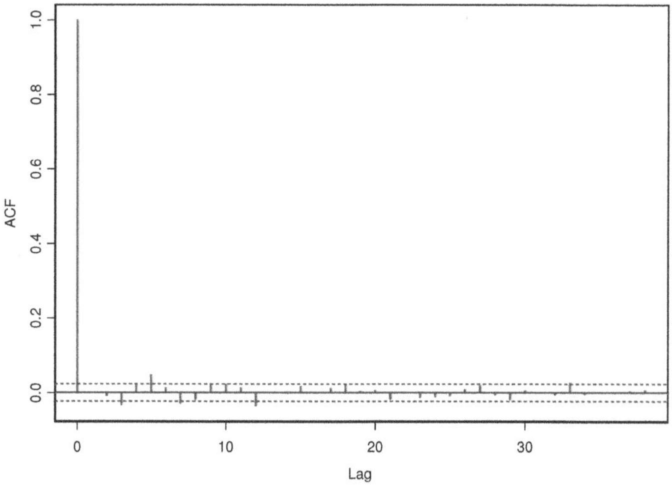

Figure 5.20: Residuals of the fitted TARFIMA(100) model for the `UNC02bin1s` series: SACF.

Figure 5.21 shows the periodogram of the residuals of the fitted AR(18) model for the `UNC02bin1s` series and Figure 5.22 shows the corresponding SACF.

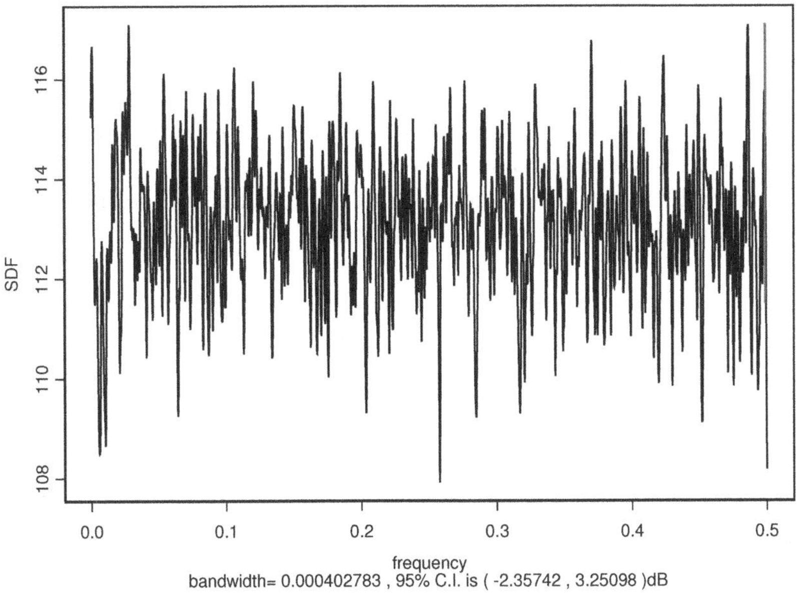

Figure 5.21: Residuals of AR(18) model: periodogram.

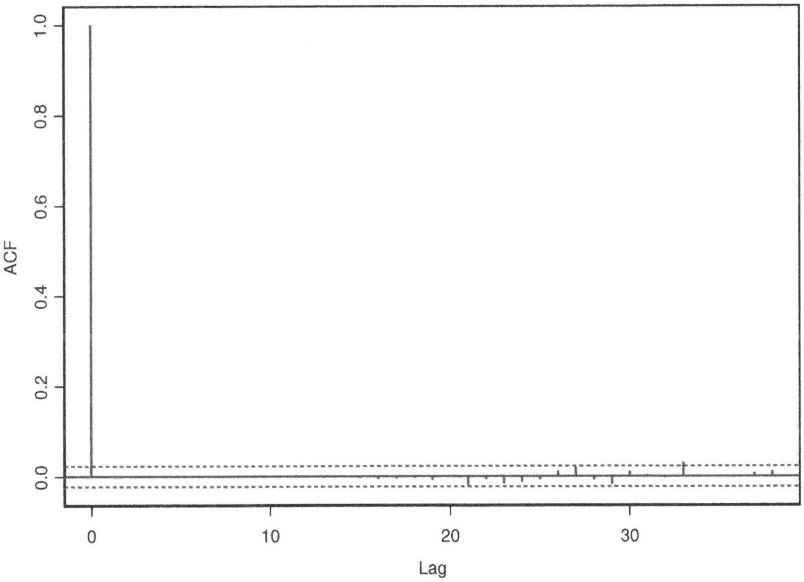

Figure 5.22: Residuals of AR(18) model: SACF.

The periodogram is practically flat and the SACF is different from zero only for the lag equal to zero. These are characteristics of a WGN indicating that the fitted AR(18) is a good model for the `UNC02bin1s` series. However, it is important to realize that the number of parameters to be estimated is 18 and the model is not parsimonious.

Figure 5.23 shows the periodogram of the residuals of the fitted AR(2) model for the `UNC02bin1s` series and Figure 5.24 shows the corresponding SACF.

It is clearly seen that the AR(2) does not perform well as the residuals do not correspond to those of a WGN: the periodogram is quite far from flat and the SACF presents non-zero values for various lags different from zero.

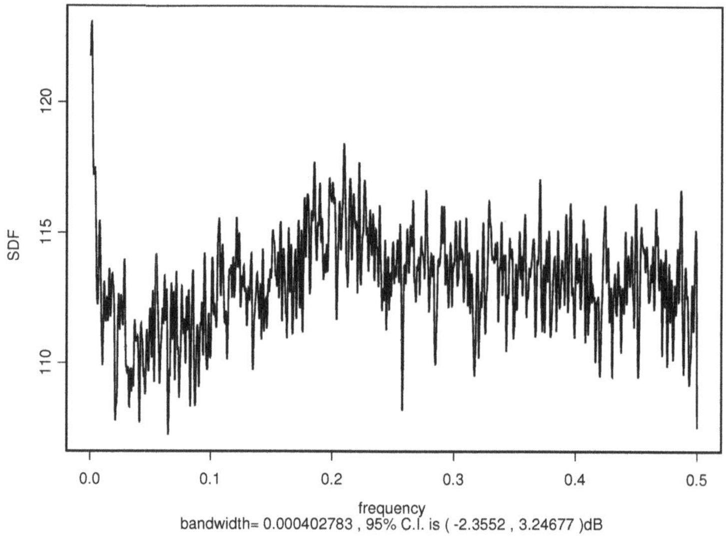

Figure 5.23: Residuals of AR(2) model: periodogram.

Figure 5.25 presents the poles and zeros diagram of the AR(2) model. In this case, the poles are inside the unit radius circle but the model has a poor performance.

In summary, the ARFIMA(0; 0.3717; 1), the AR(18) and the TARFIMA(100) models present similar performance. However, it should be emphasized that the TARFIMA(100) is much more parsimonious than the AR(18) (only two parameters against 18) and encourages the use of ARFIMA models in practice.

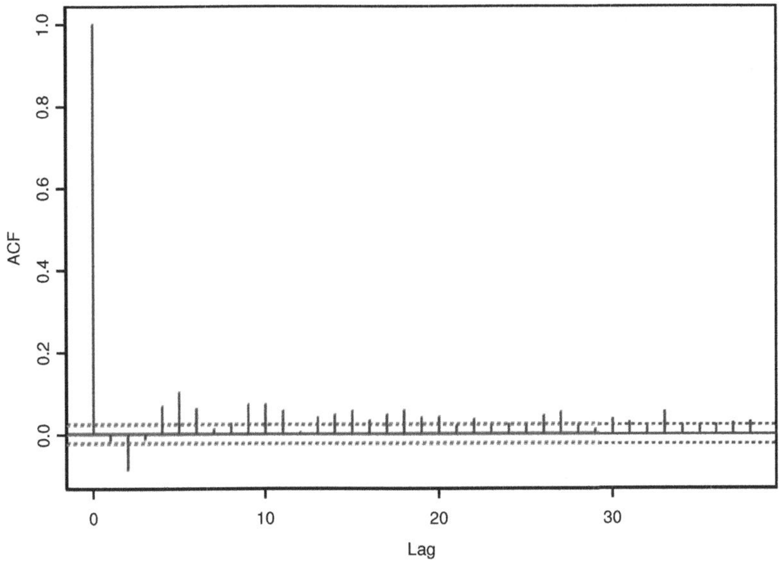

Figure 5.24: Residuals of AR(2) model: SACF.

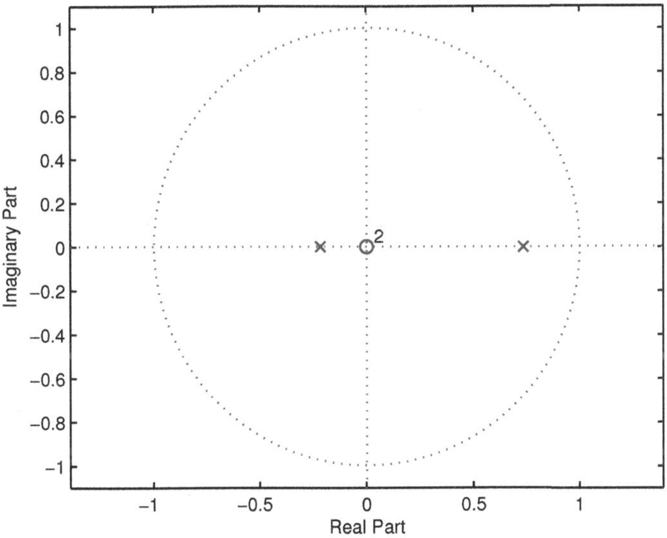

Figure 5.25: Poles and zeros diagram of the AR(2) model.

5.2.2 Long memory local analysis of the UNC02 trace

In practice, the Hurst parameter estimation methods may be severely affected by existing non-stationarities in a particular series, as sudden changes of its average level. According to the literature, this kind of regime change tends to provoke an overestimate of the H parameter H ($\hat{H} \geq 1$) made by the Abry-Veitch wavelet method (therefore, an estimate $\hat{H} \geq 1$ is an indication of presence of non-stationarity). So, the frequently adopted assumption that LRD traffic signals may be modeled by means of a stationary process characterized by a *global* H parameter, may not be realistic for many practical cases.

Figures 5.26 and 5.27 show the local analysis of the Hurst parameter of the `UNC02bin1s` series considering 256 and 512 points windows, respectively.

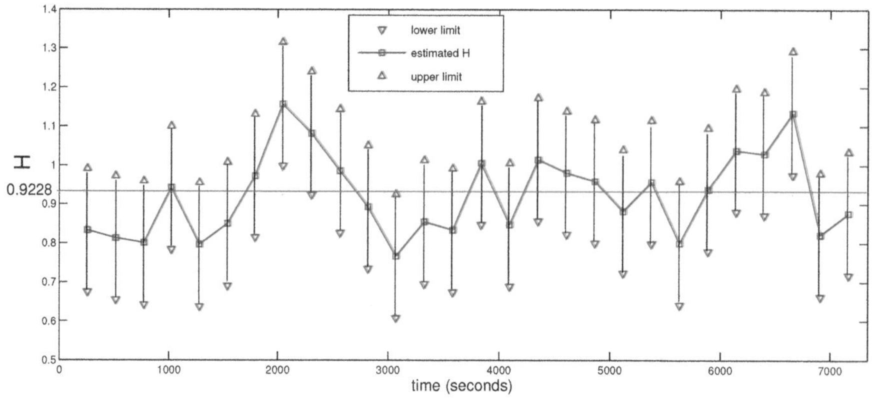

Figure 5.26: Local analysis of the Hurst parameter of the `UNC02bin1s` series: 256 points window.

The local analysis of the Hurst parameter is *blind* because:

- a) it is not known, a priori, the generator model change points location and
- b) there is not a criterium for choosing the data window's size.

However, this *empirical analysis* suggests that the windowing with a relatively small number of samples (as those made with windows of size 256) may provoke an increase of the estimator's variance, what is undesirable in practical applications.

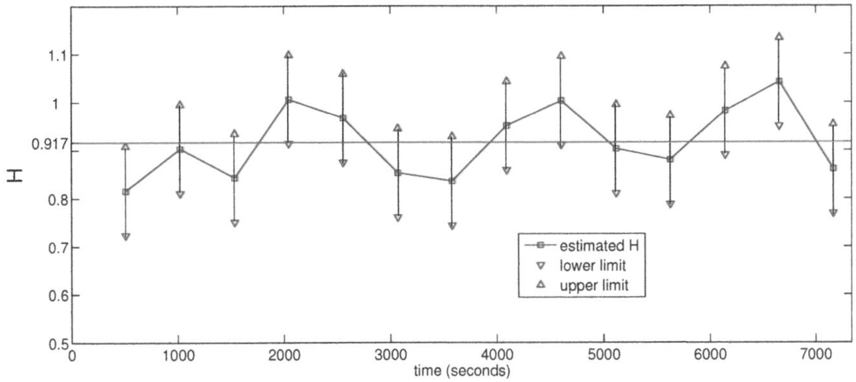

Figure 5.27: Local analysis of the Hurst parameter of the `UNC02bin1s` series: 512 points window.

5.2.3 Empirical prediction with the TARFIMA model

Figures 5.28 and 5.29 illustrate the predictions for the `UNC02bin1s` series with origin at $t = 2700$ that were obtained by means of the TARFIMA(100) and AR(18) models, and the TARFIMA(100) and AR(2) models.

- The TARFIMA(100) model used the estimate (conservative) $\hat{d} = 0.3717$.
- The AR(18) process is the best model according to the AIC criterium.
- The AR(2) and AR(18) models were estimated by the function `S-PLUS`® `AR` (Yule-Walker method).

In both cases, it is possible to see that the AR models converge too fast to the conditional mean and are not able to capture the LRD behavior of the trace.

Figure 5.30 shows the difference between the absolute prediction errors of the AR(18) and TARFIMA(100) models. Figure 5.31 shows the difference between the absolute prediction errors of the AR(2) and TARFIMA(100) models. Both figures show positive values more frequently than negative values. This means that the errors of the AR models are larger than those obtained by the TARFIMA model.

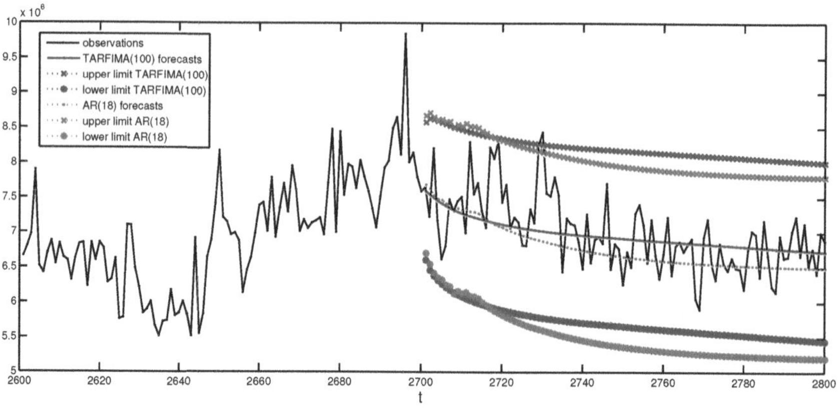

Figure 5.28: Prediction of the `UNC02bin1s` series with origin at $t = 2700$ ($\hat{d} = 0.3717$): h-step predictions and 95% confidence intervals for the TARFIMA(100) and AR(18) models' predictions.

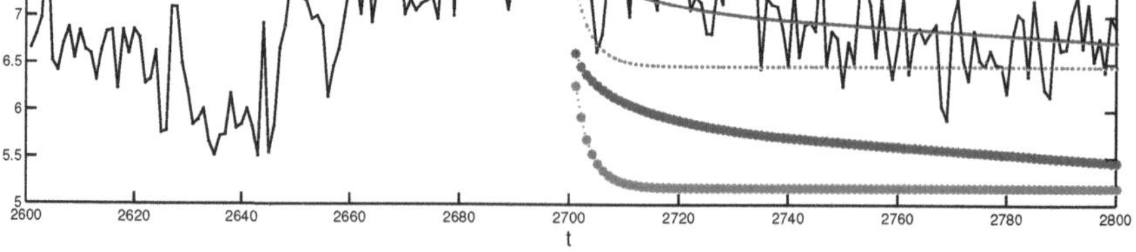

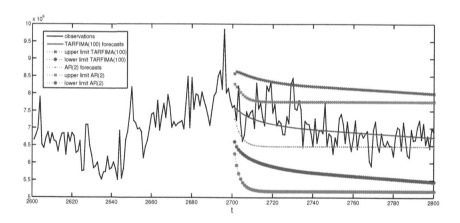

Figure 5.29: Prediction of the `UNC02bin1s` series with origin at $t = 2700$ ($\hat{d} = 0.3717$): h-step predictions and 95% confidence intervals for the TARFIMA(100) and AR(2) models' predictions.

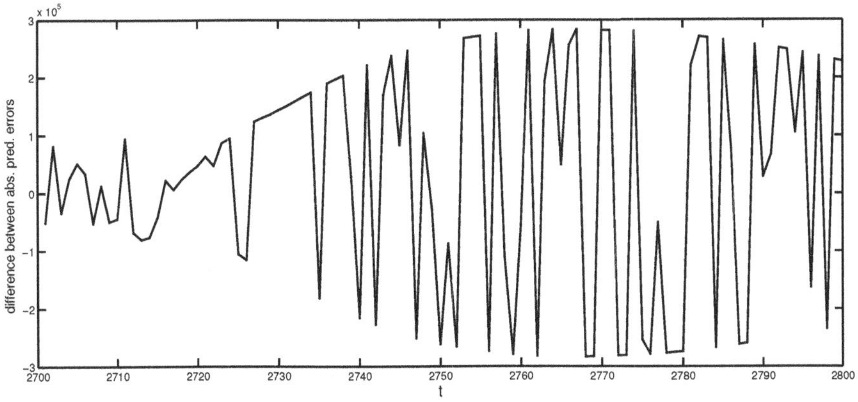

Figure 5.30: Prediction of the `UNC02bin1s` series with origin at $t = 2700$ ($\hat{d} = 0.3717$): difference between the absolute prediction errors of the AR(18) and TARFIMA(100) models.

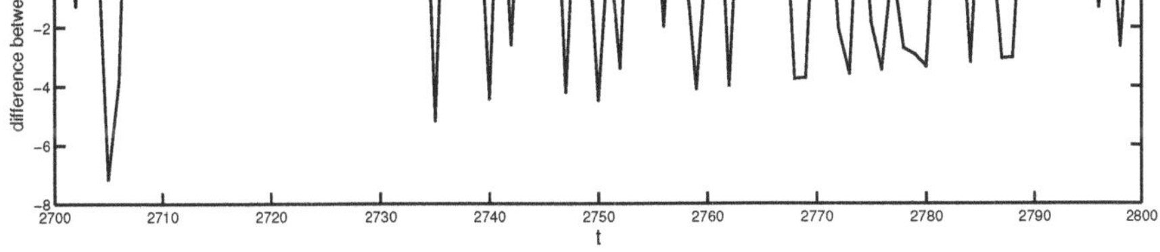

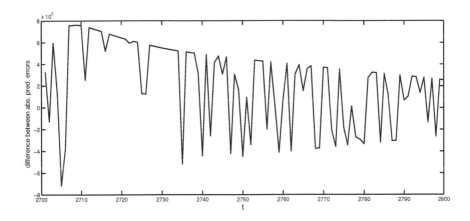

Figure 5.31: Prediction of the `UNC02bin1s` series with origin at $t = 2700$ ($\hat{d} = 0.3717$): difference between the absolute prediction errors of the AR(2) and TARFIMA(100) models.

To conclude the analysis, Table 5.3 presents the prediction power of models in relation to the AR(2) model. Negative values indicate better prediction power.

Table 5.3: `UNC02bin1s` series ($\hat{d} = 0.3717$): $PCEMSE_{x,y}(t = 2700, K)$ of the estimated AR(18), TARFIMA(10), TARFIMA(50) and TARFIMA(100) models relative to the AR(2) model.

K	AR(18)	L=10	L=50	L=100
7	-0.3933	-0.4892	-0.4365	-0.4365
10	-0.6611	-0.5307	-0.6778	-0.6778
20	-0.6885	-0.2885	-0.6858	-0.6912
50	-0.6160	-0.1791	-0.6532	-0.6655
100	-0.5448	-0.1474	-0.5989	-0.6082

It may be observed that all models perform better than the AR(2). For long range prediction, $K = (50, 100)$, the TARFIMA(50) and TARFIMA(100) models outperform the AR(18). This is a very important conclusion as the TARFIMA models are much more parsimonious than the AR(p) models with large values of p. In addition, the TARFIMA models have a direct implementation in practice by means of Kalman filters.

6

Conclusions

This chapter summarizes the contents of book and indicates some future research paths.

The purpose of this book has been to introduce the new time series TARFIMA model which has the following characteristics:

- it is a finite dimensional approximation of the ARFIMA model;
- it performs at least as well as high-order ARMA models;
- it can be implemented in practice by means of Kalman filters.

The book presents a practical review of stochastic processes, discusses in detail the Wavelet Transform and introduces several methods to determine the LRD Hurst parameter. In this way, the book provides a solid theoretical background for those interested on understanding, modeling and estimating the Internet teletraffic.

From this point on, at least two research paths are foreseen:

- it is necessary to translate the results obtained by the methods discussed in the book into performance figures understandable by the Internet community. In other words, the methods need to be used in network simulators and the achievable QoS for different scenarios evaluated in terms throughput, packet loss rate, packet delay transfer and packet delay variation;
- it is necessary to evaluate the sensitivity of the methods to errors due to finite precision implementations.

The appointed research paths are being pursued and as soon as new results become available, new publications will be produced.

So far, the techniques presented in this book have shown their value to develop QoS aware automatic network resources assignment algorithms.

Bibliography

[1] P. Abry and D. Veitch. Wavelet analysis of long-range dependent traffic. *IEEE Transactions on Information Theory*, 4(1):2–15, 1998.

[2] H. Akaike. Information theory and an extension of the maximum likelihood principle. In B. N. Petrov and F. Csaki, editors, *2nd International Symposium on Information Theory.*, pages 267–281. Akademia Kiado, Budapest, 1973.

[3] H. Akaike. A new look at the statistical model identification. *IEEE Transactions on Automatic Control*, AC-19:716–723, 1974.

[4] A. T. Andersen and B. F. Nielsen. An application of superpositions of two state Markovian sources to the modeling of self-similar behavior. In *Sixteenth Annual Joint Conf. of the IEEE Comp. and Comm. Societies, INFOCOM 1997*, pages 196–204, 1997.

[5] G. Armitage. *Quality of Service in IP Networks.* MacMillan Technical Publishing, 2000.

[6] J.-A. Bäckar. *A Framework for Implementing Fractal Traffic Models in Real Time.* Master thesis, SERC, Melbourne, 2000.

[7] Jean-Marc Bardet, Gabriel Lang, Georges Oppenheim, Anne Philippe, Stilian Stoev, and Murad S. Taqqu. Semi-parametric estimation of the long-range dependence parameter. In P. Doukan, G. Oppenheim, and M. S. Taqqu, editors, *Theory and Applications of Long Range Dependence.* Birkhäuser, Boston, MA, 2003.

[8] Stephen Bates and S. McLaughlin. Testing the gaussian assumption for self-similar teletraffic models. In *IEEE Signal Processing Workshop on Higher-Order Statistics, 1997*, pages 444–447, 1997.

[9] J. Beran. *Statistics for Long-Memory Processes.* Chapman & Hall, 1994.

[10] J. Beran, R. Sherman, M. S. Taqqu, and W. Willinger. Long-range dependence in variable-bit-rate video traffic. *IEEE Trans. Commun.*, 43(234):1566–1579, Feb./March/Apr. 1995.

[11] J. Berger and Benoit B. Mandelbrot. A new model for error clustering in telephone circuits. *IBM J. Res. Develop.*, pages 224–236, 1963.

[12] R. J. Bhansali and P. S. Kokoszka. Prediction of long-memory time series. In P. Doukan, G. Oppenheim, and M. S. Taqqu, editors, *Theory and Applications of Long Range Dependence.* Birkhäuser, Boston, MA, 2003.

[13] S. Blake. Internet RFC 2475: An Architecture for Differentiated Services, 1998.

[14] M. Bouvet and S. C. Schwartz. Comparison of adaptive and robust receivers for signal detection in ambient underwater noise. *IEEE Trans. Acoust., Speech, Signal Processing*, 37:621–626, 1989.

[15] G. E. P. Box and G. M. Jenkins. *Time Series Analysis, Forecasting and Control.* Holden Day, San Francisco, revised edition, 1976.

[16] G. E. P. Box, G. M. Jenkins, and G. C. Reinsel. *Time Series Analysis: Forecasting and Control*. Prentice Hall, 3rd edition, 1994.

[17] R. Braden, D. Clark, and S. Shenker. Internet RFC 1633: Integrated Services in the Internet Architecture, 1994.

[18] D. R. Brillinger. An introduction to polyspectra. *Ann. Math. Statist.*, 36:1351–1374, 1965.

[19] P J. Brockwell and R. A. Davis. *Introduction to Time Series and Forecasting*. Springer-Verlag, New York, 1996.

[20] Peter J. Brockwell and Richard A. Davis. *Time series: theory and methods*. Springer-Verlag, New York, 2 edition, 1991.

[21] Georg Cantor. über unendliche, lineare punktmannigfaltigkeiten v. *Mathematische Annalen*, 21:545–591, 1883.

[22] P.-R. Chang and J.-T. Hu. Optimal nonlinear adaptive prediction and modeling of MPEG video in ATM networks using pipelined recurrent neural networks. *IEEE J. Select. Areas Commun.*, 15(8):1087–1100, 1997.

[23] M. Corradi, R. G. Garroppo, S.Giordano, and M. Pagano. Analysis of FARIMA processes in the modeling of broadband traffic. In *International Conference on Communications 2000* (ICC'01), volume 3, pages 964–968, 2001.

[24] M. E. Crovella and A. Bestavros. Self-similarity in world wide web traffic - evidence and possible causes. In *Proceedings of the ACM Sigmetrics'96*, pages 160–169, 1996.

[25] I. Daubechies. *Ten Lectures on Wavelets*. SIAM, Philadelphia, 1992.

[26] Ingrid Daubechies. Orthonormal bases of compactly supported wavelets. *Comm. Pure Appl. Math.*, 41:909–996, 1988.

[27] Bruce S. Davie and Yakov Rekhter. *MPLS: Technology and Applications*. Morgan Kaufmann, first edition, May 2000.

[28] A. B. de Lima, M. Lipas, F. L. de Mello, and J. R. de A. Amazonas. A generator of teletraffic with long and short-range dependence. In *12th Computer Aided Modeling and Design of Communication Links and Networks (CAMAD07) Workshop, part of the 18th Annual IEEE International Symposium on Personal, Indoor and Mobile Radio Communications (PIMRC07)*, Athens, Greece, September 2007.

[29] Alexandre Barbosa de Lima. *Proposta de uma Estratégia para Controle de Admissão de Conexões Baseado em Medições de Tráfego Agregado e Caracterização de Redes IP*. Dissertação de mestrado, Escola Politécnica da USP, São Paulo, 2002.

[30] Alexandre Barbosa de Lima and José Roberto de Almeida Amazonas. State-space modeling of long-range dependent teletraffic. In *20th International Teletraffic Congress (ITC 2007)*, pages 260–271, Ottawa, Canada, June 2007.

[31] Fernando Lemos de Mello, Alexandre Barbosa de Lima, Marcelo Lipas, and José Roberto de Almeida Amazonas. Geração de séries auto-similares Gaussianas via wavelets para uso em simulações de tráfego. *IEEE Latin America Transactions*, 5(1), Março 2007.

[32] N. G. Duffield. Economies of scale for long-range dependent traffic in short buffers. *Telecommunication Systems*, 7(1-3):267–280, March 1997.

[33] J. Durbin and S. J. Koopman. *Time Series Analysis by State Space Models*. Oxford University Press, 2001.

[34] Agner Krarup Erlang. The theory of probabilities and telephone conversation. *Nyt Tidsskrift for Matematik B (first publication)*, 20, 1909.

[35] A. Erramilli, O. Narayan, and W. Willinger. Experimental queueing analysis with long-range dependent traffic. *IEEE/ACM Transactions on Networking*, 4:209–223, April 1996.

[36] A. Feldmann, A. C. Gilbert, and W. Willinger. Data networks as cascades: investigating the multifractal nature of internet WAN traffic. *Computer Communication review*, 28(4):42–55, 1998.

[37] W. Feller. *An Introduction to Probability Theory and Its Applications, Vol. 2.* John Wiley, New York, 2nd edition, 1971.

[38] Michael Frame, Benoit B. Mandelbrot, and Nial Neger. Natural fractals, class on fractal geometry offered by Yale University, 2006.

[39] D. Gabor. Theory of communication. *J. Inst. Eletr. Eng.*, 93(III):429–457, 1946.

[40] R. G. Garroppo, S. Giordano, S. Porcarelli, and G. Procissi. Testing α-stable processes in modeling broadband teletraffic. In *IEEE International Conference on Communications, 2000 (*ICC 2000*)*, volume 3, pages 1615–1619, June 2000.

[41] Ramazan Gençay, Faruk Selçuk, and Brandon Whitcher. *An Introduction to Wavelets and Other Filtering Methods in Finance and Economics.* Academic Press, 2001.

[42] J. Geweke and S. Porter-Hudak. The estimation and application of long memory time series models. *Journal of Time Series Analysis*, 4:221–237, 1983.

[43] A. C. Gilbert, W. Willinger, and A. Feldmann. Scaling analysis of conservative cascades, with applications to network traffic. *IEEE Transactions on Information Theory*, 45(3):971–991, April 1999. Special Issue on “Multiscale Statistical Signal Analysis and its Applications”.

[44] W-B. Gong, Y. Liu, V. Misra, and D. Townsley. Self-similarity and long range dependence on the internet: a second look at the evidence, origins and implications. *Computer Networks*, 48:377–399, 2005.

[45] C. W. J. Granger and R. Joyeux. An introduction to long-memory time series models and fractional differencing. *Journal of Time Series Analysis*, 1:15–29, Oct. 1980.

[46] Matthias Grossglauser and Jean-Chrysostome Bolot. On the relevance of long-range dependence in network traffic. *IEEE/ACM Transactions on Networking*, 7(5):629–640, October 1999.

[47] A. Grossmann and J. Morlet. Decomposition of hardy functions into square integrable wavelets of constant shape. *SIAM J. Math.*, 15:723–736, 1984.

[48] John A. Gubner. *Probability and Random Processes for Electrical and Computer Engineers.* Cambridge University Press, Cambridge, UK, 2006.

[49] J. D. Hamilton. *Time Series Analysis.* Princeton University Press, New Jersey, U.S.A., 1994.

[50] A. C. Harvey. *Time Series Models.* MIT Press, second edition, 1993.

[51] J. Haslett and A. E. Raftery. Space-time modelling with long-memory dependence: Assessing Ireland’s wind power resource. *Journal of Royal Statistical Society Series C*, 38:1–21, 1989.

[52] M. J. Hinich. Testing for gaussianity and linearity of a stationary time series. *J. Time Series Analysis*, 3:169–176, 1982.

[53] J. R. M. Hosking. Fractional differencing. *Biometrika*, 68:165–176, Oct. 1981.

[54] H. E. Hurst. Long-term storage capacity of reservoirs. *Trans. Am. Soc. Civil Engineers*, 116:770–799, 1951.

[55] J. Ilow. Forecasting network traffic using FARIMA models with heavy tailed innovations. In *International Conference on Acoustic, Speech and Signal Processing 2000* (ICASSP'00), volume 6, pages 3814–3817, 2000.

[56] Van Jacobson, Craig Leres, and Steven McCanne. TCPDUMP public repository.

[57] Raj Jain. *The Art of Computer Systems Performance Analysis - Techniques for Experimental Design, Simulation, and Modeling*. John Wiley & Sons, Inc., 1991.

[58] C. M. Jarque and A. K. Bera. A test for normality of observations and regression residuals. *International Statistical Review*, 55(5):163–172, 1987.

[59] G. Kaiser. *A Friendly Guide to Wavelets*. Birkhäuser, Boston, Mass., 1994.

[60] R. E. Kalman. A new approach to linear filtering and prediction problems. *Trans. ASME, J. Basic Eng.*, 82:35–45, 1960.

[61] A. Karasaridis and D. Hatzinakos. On the modeling of network traffic and fast simulation of rare events using stable self-similar processes. In *IEEE Signal Processing Workshop on Higher-Order Statistics (HOS)*, pages 268–272, July 1997.

[62] Anestis Karasaridis and Dimitrios Hatzinakos. Network heavy traffic modeling using α-stable self-similar processes. *IEEE Transactions on Communications*, 49(7):1203–1214, July 2001.

[63] S. Kim, J. Y. Lee, and D. K. Sung. A shifted gamma distribution model for long-range dependent internet traffic. *IEEE Communications Letters*, 7(3):124–126, March 2003.

[64] L. Kleinrock. *Queueing Systems*. John Wiley & Sons, 1975.

[65] Andreas Klimke. Mandelbrot set gui (for MATLAB), 2003.

[66] Piotr Kokoszka and Remigijus Leipus. Detection and estimation of changes in regime. In P. Doukan, G. Oppenheim, and M. S. Taqqu, editors, *Theory and Applications of Long Range Dependence*, pages 325–337. Birkhäuser, Boston, MA, 2003.

[67] A. Kolarov, A. Atai, and J. Hui. Application of Kalman filter in high-speed networks. In *Global Telecommunications Conference 1994* (GLOBECOM'94), volume 1, pages 624–628, 1994.

[68] D. Kwiatkowski, P. C. B. Phillips, P. Schmidt, and Y. Shin. Testing the null hypothesis of stationarity against the alternative of a unit root. *Journal of Econometrics*, 54:159–178, 1992.

[69] W. Leland, M. Taqqu, W. Willinger, and D. Wilson. On the self-similar nature of Ethernet traffic (extended version). *IEEE/ACM Transactions on Networking*, 2(1):1–15, Feb. 1994.

[70] Paul Lévy. *Calcul des Probabilités*. Gauthier-Villars, Paris, 1925.

[71] A. O. Lim and K. Ab-Hamid. Kalman predictor method for congestion avoidance in ATM networks. In TENCON 2000, pages I346–I351, 2000.

[72] A. W. Lo. Long term memory in stock market prices. *Econometrica*, 59(5):1279–1313, 1991.

[73] Sheng Ma and Chuanyi Ji. Modeling heterogeneous network traffic in wavelet domain. *IEEE Transactions on Networking*, 9(5):634–649, Oct. 2001.

[74] S. Mallat. *A Wavelet Tour of Signal Processing*. Academic Press, second edition, 1999.

[75] S. G. Mallat. Multifrequency channel decompositions of images and wavelet models. *IEEE Transactions on Acoustics, Speech, and Signal Processing*, 37:2091–2110, 1989.

[76] S. G. Mallat. Multiresolution approximations and wavelet orthonormal bases of $l^2(\mathbb{R})$. *Transactions of the American Mathematical Society*, 315:69–87, 1989.

[77] S. G. Mallat. A theory for multiresolution signal decomposition: The wavelet representation. *IEEE Transactions on Pattern Analysis and Machine Intelligence*, 11:674–693, 1989.

[78] B. B. Mandelbrot. *The Fractal Geometry of Nature*. WH Freeman, New York, 1977.

[79] B. B. Mandelbrot and J. V. Ness. Fractional brownian motions, fractional noises and applications. *SIAM Rev.*, 10:422–437, Feb. 1968.

[80] Benoit B. Mandelbrot. The variation of certain speculative prices. *J. Business*, 36:394–419, 1963.

[81] D. Middleton. Statistical physical models of electromagnetic interference. *IEEE Trans. Electromag. Compat.*, EMC-19:106–127, 1977.

[82] Ina Minei and Julian Lucek. *MPLS-Enabled Applications: Emerging Developments and New Technologies*. John Wiley & Sons, October 2005.

[83] P. A. Morettin and C. M. C. Toloi. *Análise de Séries Temporais*. Edgard Blücher ltda., São Paulo, SP, 2004.

[84] Pedro A. Morettin. *Econometria Financeira: Um curso em Séries Temporais Financeiras*. 2003.

[85] Pedro A. Morettin. *Econometria Financeira*. Associação Brasileira de Estatística, 2006.

[86] K. Nichols. Internet RFC 2474: Definition of the Differentiated Services Field (DS field) in the IPv4 and IPv6 Headers, 1998.

[87] Chrysostomos L. Nikias and Athina P. Petropulu. *Higher-order spectra analysis: a nonlinear signal processing framework*. Prentice-Hall, Englewood Cliffs, NJ, 1993.

[88] John Nolan. An introduction to stable distributions.

[89] Alan V. Oppenheim and Ronald W. Schafer. *Discrete-Time Signal Processing*. Prentice-Hall International,Inc., N.J., U.S.A., 1989.

[90] Athanasios Papoulis. *Probability, Random Variables, and Stochastic Processes*. McGraw-Hill, third edition, 1996.

[91] C. Park, F. Hernandez-Campos, J. S. Marron, and F. D. Smith. Trace UNC02.

[92] C. Park, F. Hernandez-Campos, J. S. Marron, and F. D. Smith. Long-range dependence in a changing internet traffic mix. *Computer Networks*, 48:401–422, 2005.

[93] V. Paxson. Fast, approximate synthesis of fractional Gaussian noise for generating self-similar network traffic. *Computer Communication review*, 27:5–18, Oct. 1997.

[94] V. Paxson and S. Floyd. Wide-area traffic: The failure of Poisson modeling. *IEEE/ACM Transactions on Networking*, 3(3):226–244, June 1995.

[95] Heinz-Otto Peitgen, Hartmut Jürgens, and Dietmar Saupe. *Chaos and Fractals: New Frontiers of Science*. Springer-Verlag, 1992.

[96] D. B. Percival and A. T. Walden. *Spectral Analysis for Physical Applications*. Cambridge, New York, 1993.

[97] D. B. Percival and A. T. Walden. *Wavelet Methods for Time Series Analysis*. Cambridge University Press, 2000.

[98] A. P. Petropulu, J.-C. Pesquet, X. Yang, and J. Yin. Power-law shot noise and relationship to long-memory processes. *IEEE Trans. Signal Processing*, 48(7), July 1989.

[99] Athina P. Petropulu and Xueshi Yang. Data traffic modeling - a signal processing perspective. In Kenneth E. Barner and Gonzalo R. Arce, editors, *Nonlinear Signal and Image Processing - Theory, Methods, and Applications*. CRC Press, 2004.

[100] R. D. Pierce. Application of the positive alpha-stable distribution. In *IEEE Signal Processing Workshop on Higher-Order Statistics*, pages 420–424, Alberta, Canada, July 1997.

[101] S. S. Pillai and M. Harisankar. Simulated performance of a DS spread spectrum system in impulsive atmospheric noise. *IEEE Trans. Electromag. Compat.*, EMC-29:80–82, 1987.

[102] John G. Proakis and Dimitris G. Manolakis. *Digital Signal Processing*. Pearson Prentice Hall, fourth edition, 2007.

[103] V. Ribeiro, R. Riedi, M. S. Crouse, and R. G. Baraniuk. Simulation of nonGaussian long-range dependent traffic using wavelets. In *SIGMETRICS'99*, May 1999.

[104] R. Riedi and J. L. Véhel. Multifractal properties of tcp traffic: a numerical study, tech. rep. 3129. Technical report, INRIA Rocquencourt, France, 1997.

[105] R. H. Riedi, M. S. Crouse, V. J. Ribeiro, and R. G. Baraniuk. A multifractal wavelet model with application to network traffic. *IEEE Transactions on Information Theory*, 45(3):992–1018, April 1999.

[106] E. Rosen, A. Viswanathan, and R. Callon. Internet RFC 3031: Multiprotocol Label Switching Architecture, 2001.

[107] M. Rosenblatt. *Handbook of Statistics, 3, D. R. Brillinger and P. R. Krishnaiah (Eds.)*. Elsevier Science Publishers B. V., 1983.

[108] N. Sadek, A. Khotanzad, and T. Chen. ATM dynamic bandwidth allocation using F-ARIMA prediction model. In 12th *International Conference on Computer Communications and Networks (ICCCN) 2003*, pages 359–363, Oct. 2003.

[109] G. Samorodnitsky and M. S. Taqqu. *Stable non-Gaussian random processes*. Chapman & Hall, London, UK, 1994.

[110] A. H. Sayed. *Fundamentals of Adaptive Filtering*. John Wiley & Sons, Inc., Hoboken, NJ, 2003.

[111] Min Shao and C. Nikias. Signal processing with fractional lower order moments: Stable processes and their applications. *Proceedings of the IEEE*, 81(7), July 1993.

[112] Y. Shu, Z. Jin, L. Zhang, L. Wang, and O. W. W. Yang. Traffic prediction using FARIMA models. In *International Conference on Communications 1999* (ICC'99), volume 2, pages 891–895, 1999.

[113] Y. Shu, Z. Jin, L. Zhang, L. Wang, and O. W. W. Yang. Prediction-based admission control using FARIMA models. In *International Conference on Communications 2000* (ICC'00), volume 2, pages 1325–1329, 2000.

[114] H. Stark and J. W. Woods. *Probability and Random Processes with Applications to Signal Processing*. Prentice Hall, Upper Saddle River, NY, 3rd edition, 2002.

[115] S. Stoev, M. S. Taqqu, C. Park, and J. S. Marron. On the wavelet spectrum diagnostic for Hurst parameter estimation in the analysis of internet traffic. *Computer Networks*, 48:423–444, 2005.

[116] M. Taqqu, V. Teverovsky, and W. Willinger. Estimators for long-range dependence: An empirical study. *Fractals*, 3:785–798, 1995.

[117] Murad S. Taqqu. Fractional brownian motion and long-range dependence. In P. Doukan, G. Oppenheim, and M. S. Taqqu, editors, *Theory and Applications of Long Range Dependence*. Birkhäuser, Boston, MA, 2003.

[118] Ruey S. Tsay. *Analysis of Financial Time Series*. John Wiley and Sons, Hoboken, New Jersey, second edition, 2005.

[119] B. Tsybakov and N. D. Georganas. On self-similar traffic in ATM queues: definitions, overflow probability bound, and cell delay distribution. *IEEE Trans. Networking*, 5:397–409, June 1997.

[120] D. Veitch, M. S. Taqqu, and P. Abry. Meaningful MRA initialization for discrete time series. *Signal Processing*, 80:1971–1983, 2000.

[121] Zheng Wang. *Internet QoS: Architectures and Mechanisms for Quality of Service*. Morgan Kaufmann, first edition, March 2001.

[122] Brandon J. Whitcher. *Assessing Nonstationary Time Series Using Wavelets*. Doctoral thesis, University of Washington, 1998.

[123] W. Willinger, R. Govindan, S. Jamin, V. Paxson, and S. Shenker. Scaling phenomena in the internet: Critically examining critically. In *National Academy of Sciences, Colloquium*, volume 99, pages 2573–2580, Feb. 2002.

[124] W. Willinger, M. S. Taqqu, R. Sherman, and D. V. Wilson. Self-similarity through high-variability: statistical analysis of Ethernet LAN traffic at the source level. *IEEE/ACM Trans. on Networking*, 5:71–86, 1997.

[125] Walter Willinger, Daniel V. Wilson, Will E. Leland, and M. Taqqu. On traffic measurements that defy traffic models (and vice-versa): Self-similar traffic modeling for high-speed networks. *ConneXions*, 8(11):14–24, 1994.

[126] X. Yang and A. P. Petropulu. The extended alternating fractal renewal process for modeling traffic in high-speed communications networks. *IEEE Transactions on Signal Processing*, 49(7):1349–1363, July 2001.

[127] F. Zhinjun, Z. Yuanhua, and Z. Daowen. Kalman optimized model for MPEG-4 VBR sources. *IEEE Transactions on Consumer Electronics*, 50(2):688–690, May 2004.

[128] Eric Zivot and Jiahui Wang. *Modeling Financial Time Series with S-PLUS*. Springer, 2003.

Index

$1/f$ noise, 30
α-stable distribution, 14, 38, 129
DFBM process, 36

Fourier Transform - FT, 50

Abry and Veitch's wavelet estimator, 80
aggregate process, 35
Akaike - AIC, 20
ARFIMA model, 30, 69, 114
ARFIMA model
  confidence interval, 74
  prediction, 71, 75
ARIMA model, 21
ARIMA model
  representations, 22
ARMA model
  autocorrelation, 34
auto-regressive models - AR, 16
auto-regressive models - AR
  estimation, 20
  identification, 18
  truncated representation, 90
autocorrelation, 10
autocovariance, 10

Bayesian Information Criteria - BIC, 20
Bi-spectrum, 82, 86
bicoherence, 87
Box-Jenkins, 13
Brownian movement, 36

central limit, 41
co-difference, 42
co-variation, 42
confidence interval, 35
correlation coefficient, 10
Cummulative Empirical Mean Square Error - CEMSE, 116
cumulants, 85
Cumulative Distribution Function - CDF, 9

Daubechies, Ingrid, 63
DFBM process, 48
Discrete Time Fourier Transform - DFT, 15

Ergodicity, 12
Ethernet traffic, 27

FBM process, 48
FD process, 70
FGN process, 36, 48
fractal, 25

generalized central limit theorem, 15, 41
GPH test, 77, 97, 103

Haslett and Raftery's estimator, 79
heavy tail, 14
Hinich test, 97, 103

IID Process, 10
impulsiveness, 2, 131
integrator filter, 7

Kalman filter, 89, 148
Kalman filter
  prediction, 92
KPSS test, 87, 96, 103

Long Range Dependence - see LRD, 2
LRD, 2, 27, 33, 126
LRD
  α exponent, 33
  heavy tail distribution, 2, 38
  Hurst parameter, 27, 33, 144
  infinite variance, 38

Mandelbrot, Benoit, 36
Mandelbrot, Benoit B., 25
mean value, 10
Minimum Mean Squared Error - MMSE, 72
model
  AR, 16
  ARFIMA, 30, 69
  ARIMA, 21
  ARMA, 14, 34
  behavioral, 47
  diagnostic, 13
  estimation, 12
  heterogenous, 47
  homogeneous, 47
  identification, 12
  MA, 14
  MWM, 27, 66, 100
  specification, 12
  state space, 89
  structural, 47
moving average - MA, 14
MWM model, 27, 117

network traffic
  aggregate, 2
  control, 2
  engineering, 1
  Ethernet, 27
  fractal, 44
  Gigabit Ethernet, 126
  Markovian, 2
Newey-West estimator, 76
Nile river, 105
Nile river
  wavelet spectrum, 109
  ACF, 111
  QQ-plot, 109
Nolan, John P., 128
non-stationarity, 11
normalized frequency, 15

operator
  auto-regressive, 14
  backshift, 6
  delay, 5
  difference, 6
  moving average, 14
  summation, 7

Pareto distribution, 38
periodogram method, 78
Poisson process, 2, 27
polyspectrum, 86
PP-plot, 129
Prediction Cummulative Empirical Mean Square Error - PCEMSE, 117
probability density function - PDF, 9
probability distribution function, 9
Purely Stochastic Process, 9

QMF filters, 62
Quality of Experience - QoE, 2
Quality of Service - QoS, 1
quantile, 32

R/S statistics, 75, 77, 96, 103
random walk, 22

sample autocorrelation - SACF, 11
sample autocovariance, 11
sample mean, 11
sample mean variance, 31
sample variance, 12
self-similarity, 2, 25, 36, 126
self-similarity
    asymptotic second order, 37
    exact second order, 37
spectral density function - SDF, 15
SRD, 27
stochastic processes, 8
Strict Sense Stationarity, 10

TARFIMA model, 90, 114, 142
TARFIMA model
    prediction power, 93
time series, 3
time series modeling, 12
traffic - see network traffic, 1
tri-spectrum, 86

van Ness, 36
variance plot, 78

wavelet, 50
wavelet
    Coiflet, 51
    Continuous Wavelet Transform - CWT, 53
    Daubechies, 51
    dilation equation, 60
    Discrete Wavelet Transform, 54
    Gaussian, 51
    Haar, 51
    Mexican hat, 51
    Meyer, 51
    Multiresolution analysis - MRA, 55, 64
    spectrum, 81, 133
    Symmlet, 51
White Gaussian Noise - WGN, 11
White Independent Noise - WIN, 11
White Independent Noise - WIN - SDF, 15
    autocorrelation, 15
White Noise - WN, 11
Whittle's method, 79
Wide Sense Stationarity, 10
Wiener's process, 36
Windowed Fourier Transform - WFT, 50

Yule-Walker equations, 18

ZZ-plot, 129

About the Authors

Professor José Roberto de Almeida Amazonas graduated from Escola Politécnica of the University of São Paulo, Brazil. He got his MSc, PhD and Pos-PhD titles from the same institution. Currently he is associated professor of the Telecommunications and Control Engineering Department. His research interests include traffic modeling and estimation, multi-constraint routing, Bayesian multiuser detection, mobile learning and Internet of Things. He is also visiting professor at University of Antioquia, Colombia and Universidad Politécnica de Catalunia, Spain.

Professor Alexandre Barbosa de Lima graduated from Escola Politécnica of the University of São Paulo, Brazil, and from Escola Naval, Brazil. He got his MSc, and PhD titles from the same institution. Currently he is a professor of Electrical Engineering at Pontifícia Universidade Católica de São Paulo, Department of Engineering. His research interests include computer networks, teletraffic and spectral analysis using wavelets.

Lightning Source UK Ltd.
Milton Keynes UK
UKOW06n1936141214

243130UK00002B/55/P